D0049396

Genetic Twists of Fate

Genetic Twists of Fate

Stanley Fields and Mark Johnston

The MIT Press
Cambridge, Massachusetts
London, England

For information about special quantity discounts, please email special_sales@mitpress.mit.edu

This book was set in Stone Sans and Stone Serif by Toppan Best-set Premedia Limited. Printed and bound in the United States of America.

Library of Congress Cataloging-in-Publication Data
Fields, Stanley.
Genetic twists of fate / Stanley Fields and Mark Johnston.
p. cm.
Includes bibliographical references and index.
ISBN 978-0-262-01470-0 (hardcover : alk. paper) 1. Medical genetics—Popular works. 2. Human genetics—Popular works. I. Johnston, Mark, 1951– II. Title.
RB155.F54 2010
616'.042—dc22

2010006926

10 9 8 7 6 5 4 3 2 1

Contents

Preface

We all know someone whose life is profoundly affected by the genes they inherited—perhaps an aunt with early-onset Alzheimer's disease, or a cousin with cystic fibrosis, or a neighbor's child with Down syndrome, or diabetes, or muscular dystrophy. Might Grandma's failing eyesight have a genetic basis? Dad's heart disease certainly seems to be genetic: Grandpa died of a heart attack in his late fifties. The consequences of altered genes can be seen all around us.

Progress in unraveling the genetic basis of disease and behavior is also evident all around us. Hardly a day goes by without a story in a newspaper or on a website proclaiming that some gene has been found to contribute to the risk for a serious disease such as diabetes, cancer, or colitis, or a condition such as depression, alcoholism or autism, or a trait such as fearlessness, aggressiveness, or anxiety.

How, we wondered, does anyone make sense of all of this? We quickly realized that almost no one does make sense of it. It's not that the research is so complex or the public too dense to understand it. No, it's simply that most people probably haven't thought much about the workings of genetics since their high school biology course, and long ago forgot the key principles. As a consequence, the revelation that a newly identified gene is linked to an increased risk for cancer has about as much context for us as when we learn that a new political party has formed in Uzbekistan. As genetics researchers, we set out to make human genetics fathomable to all those whose tax dollars generously support this research—including, for the past two decades, the research we do in our own labs.

This book seeks to answer clearly and simply the key questions nearly everyone has about genes and genetics. Why do we resemble our parents more than any other set of parents? Does cancer run in families? Why do some genetic diseases haunt families only when both parents carry the defective gene, whereas other genetic diseases are passed on to children from only one parent? What is our personal DNA code? How much of our behavior and our risk for diseases are influenced by this DNA code? Do the genes we get from Mom and Dad influence our moods? How does the single cell that is the fertilized egg become a baby made up of trillions of cells? What is evolution by natural selection (and was Darwin really the guy who came up with the idea)? This book is not a genetics text; these chapters and the stories they tell are meant to be quickly assimilated and easily digested.

To nonscientists, science can seem an endless stream of dry technical details. Yet all of us are tantalized by stories about real people—about the trials of the rich and the famous, as well as about the tribulations of the average Joe and Jane. Many of the most bizarre aspects of the tales we tell you here hinge on the inheritance of one tiny piece of DNA.

We'll consider the fates of some celebrated people: literary luminaries, including a Nobel Prize–winning novelist driven by a heavy genetic burden, a prolific writer of science fiction whose overeating led to more than obesity, and an immigrant magazine editor who confronted a U.S. president as well as a relentless disease; media and movie celebrities, like the seductive actress who slowly lost her mind, and the widow who rose to the highest ranks of network news and used her fame to promote cancer tests; and a star athlete who faced the dual challenges of a terminal illness as well as the world's prejudice.

We'll also recount a few of the most notable cases in the annals of medicine: the story of the so-called "Bubble Boy," who waited in vain for a cure that came too late; a confused and forgetful woman whose doctor gave his name to a disease that afflicts millions of us; and a young diabetic who nearly died waiting for the first batches of insulin to become available.

And we'll describe ordinary citizens facing extraordinary challenges: a mother wrongly accused of poisoning her infant son; mountain-climbing brothers at risk for a neurological disease; a man who made sure that all his relatives gave blood in a quest to uncover the cause of their mysterious affliction.

What all these tales have in common is a twist of genetic fate: the inheritance of one minuscule change rather than another in our vast personal DNA code, a change that made the difference between health and disease, between happiness and heartbreak, between life and death.

In the course of telling these people's tales we can't help but introduce a few of our fellow biologists who have contributed to the remarkable advances in genetics. Long before "intelligent design" was in the news, two men raced each other to explain evolution. A scientist who thought outside the box tackled a disease of cattle and ended up discovering a potent rat poison that became the most common treatment for heart attacks. A physicist split the gene, and a German woman confronted scientific orthodoxy and laid bare how a baby can develop from a single cell.

We are confident that after reading these accounts, you will understand what genes are and how scientists track them down. You will be able to appreciate why particular versions of genes in a personal DNA code put someone at risk for diabetes, or dementia, or depression. And we hope that the principles of genetics presented in this book help you understand the results of a genetic test that you or a loved one may someday undergo.

Acknowledgments

To writers, the encouragement and feedback provided by readers of early drafts are like the first drops of sugary water on a packet of dried yeast—they leaven the creative process. We have benefited tremendously from several diligent and insightful readers of early versions of this manuscript, including Christine Berg, Ed Fields, Jay Lawrence, Betty Olson, and Eric Phizicky. Others who gave us valuable comments and encouragement for which we're grateful include Josh Akey, Laurie Carr, Tracey DePellegrin-Connelly, Claire Delahunty, Barb Flatoff, Bill Schwarzer, Ken Smith, Gretchen Smith, Brien Stafford, Ellen Taussig, Bruce Tidor, and, on specific chapters, Phillip Chance, Malia Fullerton, Charles Laird, Bill Sly, and Nancy Wexler.

We thank our colleagues in our laboratories, especially Jim Dover, for their dedication to research. We appreciate the support of the Howard Hughes Medical Institute, the James S. McDonnell Foundation, the National Institutes of Health, and the University of Washington in Seattle and Washington University in St. Louis.

A particular pleasure for us was the opportunity to work on this book in peaceful and beautiful locations, for which we thank the helpful folks at the Whiteley Center at Friday Harbor Laboratories of the University of Washington; Sir Greg Winter and the fellows of Trinity College, Cambridge; Grahame and Lynn Hardie and the University of Dundee, Scotland; and Susan Dutcher and Gary and Owen Stormo, who extended hospitality to us in their Brian Head, Utah, home.

Finally, we thank our wives and families for their support and good cheer throughout the time we spent working on this project.

1 Your Personal Genome: Googling Your DNA

Not many homeowners can boast having a garage that changed the world. But Susan Wojcicki can. She can look back at her decision in 1998 to rent out her garage at 232 Santa Margarita Avenue in Menlo Park, California, as a world-changing event. Her renters, two graduate students in computer science at nearby Stanford University, needed space to develop a new company around their revolutionary approach to searching the web. Sergey Brin and Larry Page would not occupy her garage for long; they soon needed more spacious headquarters for the company that would launch their combined net worth into the stratospheric level occupied by the likes of Bill Gates and Warren Buffett.

But before they launched Google, Brin and Page devised something they called PageRank, a method to rank a webpage according to how many other pages have links to it, and the number of links each of those other pages in turn has to yet other pages. PageRank provided a much more effective means of searching the Web than that offered by then-available search engines such as AltaVista or Excite. Capitalizing on its vast ability to organize and serve up information on the Web and provide a host of other services—Google Earth, Google Image, Google News, Google Maps, Google Groups, Google Books, and more—Google quickly went beyond PageRank as its instrument for dominating the lucrative search market.

Susan Wojcicki must have realized that her young tenants were on to something: she became one of Google's first employees, going on to develop its online advertising business. But in addition to connecting people to information, Wojcicki also introduced Sergey Brin to her younger sister Anne. That introduction culminated in the May 2007 wedding of Brin and Anne Wojcicki, then both thirty-three, on a private island in the

Bahamas. The wedding couple wore bathing suits, and some of the guests joined them in swimming to the ceremony.

Like her husband, Anne Wojcicki is one half of a business duo that launched a start-up company. But the company founded by Wojcicki and her partner Linda Avey—23andMe—does not deal in Internet search results, digital images, or city maps. Instead, it serves up genetic information. Despite Google's investment of $3.9 million in 23andMe, this start-up, and other similar ventures such as Navigenics, deCODEme, Knome, and Sciona, leverage no recent insight from computer science. They rely on a principle known for over seventy years: the more closely related two individuals are to each other, the more similar are their personal DNA codes. This is the principle that allows geneticists to figure out whether a particular version of a gene makes an individual who carries it more prone to a disease such as diabetes, Parkinson's disease, or cancer than does another version. This is also the guiding principle of our book, as we lead you to an understanding of what DNA is, how its four-letter alphabet spells out the genes that determine the traits of human beings with all our complexity, and how the way those four letters line up six billion times in our personal DNA code influences how we look, how we behave, how we get sick, and how we respond to treatment.

Named after the 23 pairs of chromosomes we carry in every one of our cells, 23andMe has as its mission "to take the genetic revolution to a new level by offering a secure, web-based service where individuals can explore, share and better understand their own genetic information." It can't provide an individual with her complete DNA code, but for $999 it can reveal enough of it (about 1/6000th) to give her a glimpse of her risk for a few diseases. The web-based service the company provides allows you, in effect, to Google your own DNA: instead off typing words into a search engine to scan the world's web pages, you type the DNA letters of a gene to scan the world of your own DNA, searching for the relevant text stored therein.

Today their service is at the cutting edge; in a few years it will seem primitive. This mere glimpse of a customer's DNA will soon be replaced by a complete reading of how the four letters of its alphabet are used 6 billion times to spell out his DNA code. So far, the DNA code of only a few people has been read, at a cost of over $1 million each, prohibitive for all but the wealthiest. But new technologies are rapidly reducing the cost, bringing this brave new world into view. An added bonus to the acclaim and profits that will accrue to those who solve the issues of cost and speed is the Archon X Prize of $10 million, which will be awarded to the first team that reads the DNA code of one hundred people in ten days at a cost of ten thousand dollars each. It is possible, maybe even likely, that that reward will have been claimed by the time you read these words.

While a reading of our complete DNA code may be still out of reach for you and for us, it was realized recently by James D. Watson. More than half a century ago, on February 28, 1953, Watson and his colleague, Francis Crick, launched an age of genetic discovery with their announcement to the lunchtime patrons of the Eagle Pub in Cambridge, England, that they had "found the secret of life." Their discovery of the structure of DNA—the most important molecule of life, which specifies the form and function of every living thing—made clear how traits are passed down through the generations. Watson and Crick's breakthrough paved the way for an age of discovery that culminated in the announcement on June 25, 2000—not in a pub but at the White House—that the human DNA code had been determined. A few years later Watson himself became one of the first two people to read his own personal DNA code.

After 1953, Watson went on to a celebrated career, directing a laboratory at Harvard University, then a storied scientific institution at Cold Spring Harbor on Long Island, and ultimately the Human Genome Project, which deciphered the human DNA code. But shortly before his own DNA code was determined, Watson's professional life ended amid charges of racism. He was quoted in the October 14, 2007 edition of The Sunday Times that he was "inherently gloomy about the prospect of Africa" because "all our social policies are based on the fact that their intelligence is the same as ours—whereas all the testing says not really." He also said, "There is no firm reason to anticipate that the intellectual capacities of peoples geographically separated in their evolution should prove to have evolved identically. Our wanting to reserve equal powers of reason as some universal heritage of humanity will not be enough to make it so."

As his comments rapidly circled the globe, drawing condemnation from his fellow scientists, the 1962 Nobel Laureate quickly apologized for them. Speaking at a meeting of the Royal Society in London on October 18, 2007, he said, "To all those who have drawn the inference from my words that

Africa, as a continent, is somehow genetically inferior, I can only apologize unreservedly. . . . That is not what I meant. More importantly, there is no scientific basis for such a belief."

But the damage was done, and so was Watson's job. The board of trustees of Cold Spring Harbor Laboratory relieved Watson of his position as chancellor. They wrote, "The comments attributed to Dr. James Watson that first appeared in . . . *The Sunday Times* U.K. are his own personal statements and in no way reflect the mission, goals, or principles of Cold Spring Harbor Laboratory's Board, administration or faculty. . . . The Board of Trustees, administration and faculty vehemently disagree with these statements and are bewildered and saddened if he indeed made such comments."

Watson had steered into the always-dangerous shoals of the genetics of race, and he should not have been surprised that his words sank him. In our penultimate chapter we, too, venture into these treacherous waters. We will show you that there are many more genetic differences *within* racially defined populations such as Africans and Caucasians than *between* these populations. You can see the close resemblance of the DNA codes of these races if you compare the few available sequences. Or, you can wait a few years and see it when you read your entire DNA code.

Just as Google's computers read a digital code composed of 1s and 0s, living creatures read a chemical code of four different units, abbreviated as A, C, G and T. We'll see how strings of these four chemicals get decoded in the production of proteins, the workhorses of the body that enable us to move, see, breathe, think, and reproduce. These four chemical units are strung together 6 billion times (6 followed by 9 zeros, or 6 thousand thousand thousand)—but this number is infinitesimal compared to a "googol"—1 followed by a hundred zeros, or ten thousand trillion trillion trillion trillion trillion trillion trillion trillion trillion trillion. Yet even a googol is barely a speck in comparison to a "googolplex," which is 10 raised to the power of one googol, or 1 followed by 10^{100} zeros. (It would take much more space than the pages in this book to write that number.)

The algorithm Larry Page developed to search the Web originally went by the unbusinesslike name of BackRub. But in late 1997, as he and Sergey Brin contemplated starting a company to exploit their search engine, BackRub had to go in favor of a more fashionable term that would connote the vastness of what they were trying to organize. Unfortunately, the names they first came up with had already been claimed by other people. Page's officemate Sean Anderson made a number of suggestions, but Page nixed all of them. Anderson eventually offered "Googolplex," a name that suggested the vast amount of information the new search engine could scan. Page liked it, but preferred the shorter "Googol." Computational brilliance they may have possessed, but world-class spelling was not their forte. When Anderson used the new search engine to see if the name was available, he typed in "Google" and found that it was unclaimed. That evening Page registered the domain name Google.com. Only the next day did they learn they had misspelled the term, and discovered that the domain name "Googol.com" had in fact been claimed.

As Google rapidly expanded, Brin and Page focused on maintaining its spirit of adventure and cohesiveness. Employee number 56, who arrived in November 1999, was Charlie Ayers, their executive chef. Ayers provided free, wholesome food to the young Google workforce, maintaining their energy for the ambitious tasks they were tackling. He later recalled to David A. Vise and Mark Malseed, authors of *The Google Story*, "I could feel the energy. They had it. Everyone was so focused and into it, and they all had one goal: to make this company successful. It was 'Look at what we did,' not 'Look at me.'"

An equivalent organizational spirit exists within every one of the trillions of cells in your body. DNA provides the corporate vision and hiring plan, but it's the roughly twenty thousand varieties of proteins that carry out all the necessary activities of the cell. Like Google employees, proteins engage in a team effort that is much greater than the sum of their parts. You'll see as you read on how proteins read the DNA code in the single cell that is the fertilized egg and tell it to divide into two, then four, then eight and so on. Successive generations of cells take on new functions, specializing as heart and blood, brain and nerves, bone and teeth and all the other tissues of the body. The result, a living human being, is more magnificent than any company, no matter how much revenue it generates.

Sergey Brin was born in the Soviet Union in 1973 to two mathematicians. His father, Michael, is now a professor at the University of Maryland, and his mother, Eugenia, works at NASA's Goddard Space Flight Center, in

Washington, D.C. Anti-Semitism in the Soviet Union in the 1970s prevented his parents from advancing very far in their academic careers, so they decided to apply for exit visas, even though doing that meant running the risk of becoming unemployed and ostracized by colleagues. They were fortunate to be some of the last Jews allowed to leave the Soviet Union before it broke up a decade later. Michael and Eugenia Brin left with their young son in 1979, and settled into a new life in Maryland.

Larry Page, born just a few months earlier than Brin, is the son of the late Carl Victor Page, a professor of computer science at Michigan State University who was one of the first to receive a doctorate in this field of study. His mother, Gloria, earned a master's degree in computer science and taught programming at Michigan State. Page benefited from a rich exposure to computers long before most Americans had ever seen one.

Brin and Page seem to have a knack for science and technology, like their parents. They are bright, inquisitive, and creative, like their parents. They have a risk-taking, adventuresome attitude, like their parents. Humans have recognized for thousands of years that offspring resemble their parents, knowledge that they applied early in the course of human civilization to the selective breeding of plants and animals. Yet if you could compare the strings of DNA units in Brin's or Page's personal DNA code to the strings in your own DNA code, you would find that they are 99.9 percent identical. If you lined up your DNA code with that of Tiger Woods, or Madonna, or Barack Obama, George W. Bush, Hillary Rodham Clinton, or Paris Hilton—you would find that these, too, are 99.9 percent identical. Since all these people look and act differently from each other and from you, that 0.1 percent difference between any of these people must play a big role in personal appearance and behavior. And as we'll discuss in subsequent chapters, that 0.1 percent difference also results in some of us getting cancer, or Alzheimer's disease, or having the good fortune to escape these diseases altogether.

We expect that within the next ten years the parents of a newborn will be presented with the code that is written in the strings of letters of her DNA, in addition to their child's footprints and thumbprints and APGAR score. They will be able to Google that code and predict their child's risk for some diseases and behaviors. They will know what to watch for, and in some cases they will be able to intervene to minimize unwanted consequences and maximize desirable outcomes. The child born ten years from now will have unprecedented self-awareness (genetically speaking), and unheard of self-control (medically speaking).

Clearly, we are on the cusp of a genetic revolution, one that will affect all of us in a personal way, every day. To benefit from this revolution and manage the consequences, everyone needs to understand what is meant by "personal DNA code" and how the small fraction of differences between individuals' long lists of DNA letters makes each of us unique. How does that code determine who we are, what diseases we may get, how we feel, and what we're capable of? How much control does our personal DNA code exert over our fate?

In this book we provide some answers to the questions: What is a "personal DNA code" and what will it tell us about ourselves? How do the genes specified in the human DNA code get assigned to specific cellular functions? How do individuals' differences in this code result in differences in disease risk? The answers require us first to explain what DNA and genes are, how they are inherited, and how they collaborate to allow a fertilized egg to turn into a creature of amazing complexity in nine short months. From there, we can begin to show you why each person's DNA code is a little bit different from everyone else's, and how these variations influence our lives.

I What Do Genes Do?

2 Genes Are the Instructions for Life: AIDS and the Uncommon Man

Just one small change in one gene might have given the world more books like *Pebble in the Sky, The Stars Like Dust, The Foundation Trilogy*, and *I, Robot.* To science fiction enthusiasts of a certain age, the publication of a new Isaac Asimov novel or short story was cause for celebration. Before iPods and instant messaging, before YouTube and Facebook, before Xboxes and PlayStations, young fans would curl up under the bedcovers with one of Asimov's intergalactic tales and read late into the night. For a period in the 1960s and 1970s, Asimov's large black glasses and mutton-chop sideburns made him one of the world's most recognizable authors. His books sold in the millions. Millions more might have flown off booksellers' shelves had Asimov inherited a personal DNA code with one small change.

Asimov penned more than just science fiction; he wrote on almost any topic. He explained mathematics and astronomy, chemistry and biology, as well as the Bible and Shakespeare, American history and the Roman empire, along with Gilbert and Sullivan and *Paradise Lost*, and limericks, and Egyptian history. Asimov wrote almost nonstop, averaging about a thousand words a day, every day, for fifty years. "Being a prolific writer has its disadvantages, of course," Asimov commented in one of his three autobiographies. "It complicates the writer's social and family life, for a prolific writer has to be self-absorbed . . . and has not time for anything else." This obsession for writing—and the accompanying unwillingness to do much else—had unwelcome consequences, including the breakup of his first marriage. And "a prolific writer . . . has to *love* his own writing," Asimov noted. He certainly loved writing: he wrote over four hundred books!

Asimov's many popular nonfiction works include *The Chemicals of Life* (1954), and *The Genetic Code* (1963). In the latter work he attempted to explain to a lay audience what a gene was at a time when even the research biologists working on genetics barely understood what it was.

What is a gene? Surely those research biologists must have figured it out in the forty-seven years that have intervened since Asimov weighed in on the subject. And the answer should be common knowledge to a society accustomed to headlines proclaiming "Scientists identify gene for schizophrenia," or for obesity, or for colon cancer, or for any number of other diseases and conditions.

Apparently not. None of the people we queried who weren't biologists came close to providing a definition of the gene that would pass muster in a high school biology classroom. Many associated the gene with physical traits or emotional characteristics, a reflection of the realization that genes come from our parents and grandparents, and the notion that they explain Johnny's big ears and Mary's quick temper. The gene was occasionally linked to concepts such as DNA or chromosomes, although those terms were fairly fuzzy to those more than two years out of tenth grade biology.

Most commonly, our requests to explain what a gene is were met with the same look that might surface in response to a question about what the World Bank does, or why tort reform is needed, or how tornados start. But unlike those technical details, which we can safely leave to economists, lawyers, and meteorologists, genes are too important to be relegated to the safely obscure. We fail to understand them at our peril, because they influence crucial aspects of our daily life: our health, our lifespan, our mood, our insurability, and our food supply, to name a few.

But before we deal with *what* genes are, let's ask an even simpler question: *where* in your body are genes found? This was the query posed to Americans of varied ethnic and educational backgrounds by Angela D. Lanie, a scientist at the University of Michigan. Nearly a quarter of the respondents said genes are "in the brain"; about one eighth said "in the blood"; a few said in the reproductive system, or heart, or bones, or lymph nodes or various other locations.

Only about one third of Lanie's respondents gave the correct answer: genes are present all over the body—everywhere, in virtually every cell. Genes are found in your skin and your stomach cells, your lung and your liver cells, your brain and your bone cells. They are in every part of your body. And not just in your body: they are in every cell of broccoli and beets, and chicken and cows, and apples and apricots. If you stopped eating food that contains genes, you'd have made the unpalatable (and unsustainable) choice to dine on not much more than sugar and water.

An understanding of genes is within easy reach for all of us, because the principles that govern the operation of genes are simple. What is a gene? It is a stretch of DNA that contains the instructions for the cell to manufacture a protein. Once the terms "DNA" and "protein" and "cell" have been explained, this definition is surprisingly satisfying to most nonscientists (and to most scientists, too, for that matter).

Let's start with the cell. Imagine your body as an enormous hotel composed of about 100 trillion (1 followed by fourteen zeros) rooms, each a self-contained space enclosed by a set of walls within which sits a bed, dresser, night table, and other furniture. Each cell in your body is a similarly self-contained unit measuring about a fortieth of a millimeter across (about a tenth the width of a human hair), surrounded by a flexible membrane that protects it from the environment. There are compartments in the cell where specific functions are carried out, including maintaining the cell's DNA, burning fuel to provide energy, and transporting material to where it needs to be.

Each room in a hotel has plumbing that connects it to a central water supply, a source of electricity to power its appliances, and heating and cooling units to control its temperature. A central processing system housed on the top floor of the hotel—a phone switchboard and a computer with an Internet connection—allows every room to be in contact with the front desk, and with all the other rooms in the hotel—indeed, with the rest of the world. Likewise, each cell connects to and communicates with adjacent cells and with the rest of the body using chemical and electrical signals.

From a distance the many rooms of the hotel look the same, but if we look more closely we see that some have one bed, others two. Some have separate kitchenettes, or sitting areas, or extra bedrooms. Some are so tiny that they barely accommodate one guest; others are suites that accommodate a large family. Most rooms are rectangular, but some special rooms have unusual shapes (like the octagonal honeymoon suite). Cells, too, are specialized to carry out particular functions, differing in their size and shape and internal structures. There are blood cells and brain cells, lung cells and liver cells—cells with long and narrow projections, cells with unusual talents to filter blood or detoxify alcohol.

Rooms are not arranged willy-nilly in a hotel but are arrayed in orderly wings of multiple floors. Cells are also arrayed in the body in an orderly fashion; they organize themselves into successively larger units of tissues, organs, and organ systems. Most of these systems are familiar to us. The collection of cells that make up the mouth, esophagus, stomach, small and large intestine, the liver, pancreas, and gall bladder constitutes the digestive system, which processes food. Cells of the heart, arteries, veins, and blood form the circulatory system, which delivers nutrients and oxygen to the far reaches of the body. Cells that make up the bladder and the colon cooperate to manage a storage system that holds waste until it is ready for disposal. Cells of the brain and spinal cord constitute the nervous system that manages it all.

Cells entrust the instructions for their construction and operation to DNA, a chemical found in all living things. How did we come to know that DNA serves this vital function? In 1944, Oswald T. Avery, a physician and scientist working at the Rockefeller Institute for Medical Research (now The Rockefeller University) and his coworkers Colin MacLeod and Maclyn McCarty reported that they could dramatically change the properties of a cell—in their case a cell of the bacterium that causes pneumonia—by changing only its DNA. They concluded that DNA was the long-sought substance of heredity.

Attributing such importance to DNA was a startling result, because DNA was known to be a molecule consisting of a seemingly endless, monotonous string of only a few very similar subunits. How could such a "stupid molecule" (as some then called it) determine what kind of covering enclosed a bacterial cell (the trait that Avery and his colleagues analyzed), much less perform the amazing feat in more complex creatures of specifying the appearance of limbs and lungs and livers in all the right places and of the right size, and the proper number of teeth and toes, and irises and corneas and retinas that form eyes, and much, much more? Surely, thought many biologists, a more complex molecule was needed to accomplish those amazing feats.

It's easy to see why they thought this way, because DNA is indeed a simple molecule. It is composed of only five atoms: carbon, hydrogen, oxygen, phosphorus, and nitrogen—the organic elements from which all living things are built. DNA is a polymer—a long molecule made up of

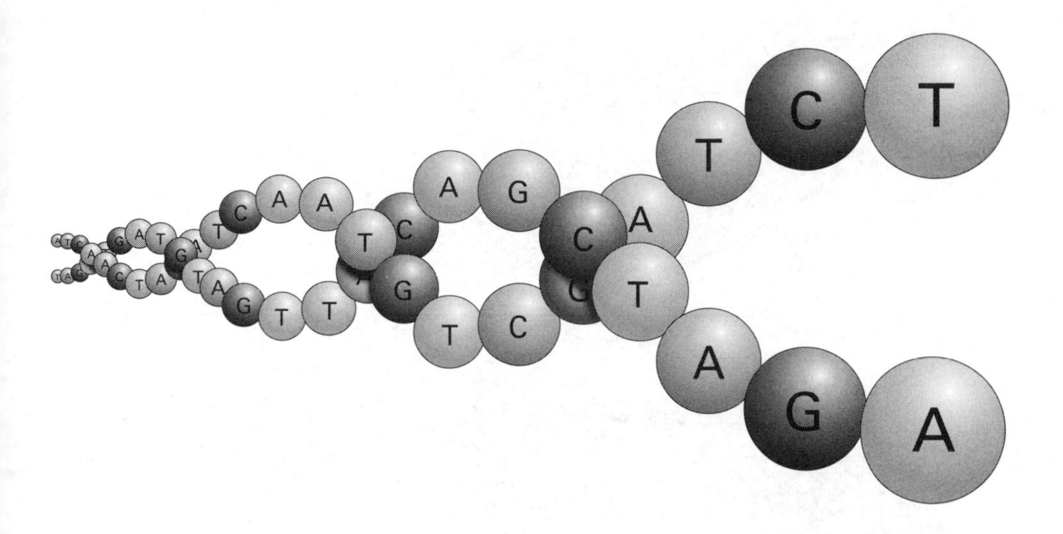

small units linked together, one after the other, like pearls in a necklace. These smaller units are known as "bases," and they come in only four types, commonly called by the first letters of their names: A, C, G and T (adenine, cytosine, guanine, and thymine).

These four bases link one to another to form a very long string—think of an extremely long necklace made up of four different kinds of pearls. Two of these strings of bases wrap around each other—think of a double pearl necklace—to form the iconic double helix structure with two strands of bases spiraling around each other (see figure).

A chromosome is a long, unbroken string of the DNA double helix (along with some packaging material to wrap up the DNA molecule so it fits inside the cell). Each human chromosome is a double string of around 100 million DNA bases. In some creatures the chromosomes may be less than one hundredth that size; in others they can be up to ten times longer.

But DNA turns out to be not so stupid a molecule, because the order of the bases—the exact sequence of the A's, C's, G's, and T's—is the information that specifies the characteristics of an organism. Of all organisms. Of us.

Now comes the most important fact: the sequence of A, C, G, and T bases needs to be specified for only one of the two strands of a DNA molecule because the sequence of bases of one of the DNA strands specifies the sequence of bases of the other. This is a result of the way the two strands of the double-helical structure are held together by interactions between the bases: bases in one strand attract and stick to bases of the other.

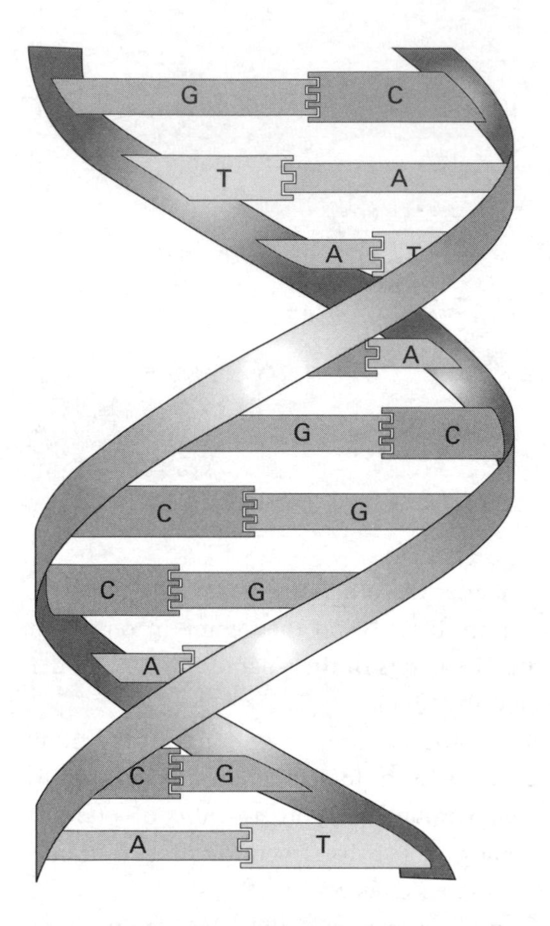

But the bases don't interact haphazardly: A and T stick to each other but not to G or C; likewise, G and C stick to each other but not to A or T. Think of 110-volt electrical plugs with two prongs and 220-volt plugs with three prongs. The bases A and T are like a matched pair of a two-pronged plug and a two-prong socket; G and C are matched like a three-pronged plug and a three-prong socket (see figure). (The "plugs" and "sockets" of the double helix are really a kind of chemical bond—a quite weak one—called a "hydrogen bond"; G and C have three of them, A and T have two.) As a consequence of these specific matches, each strand of the helix carries the information to specify its partner strand's type.

When a cell divides to form two cells, it copies the two strands by peeling them apart—unplugging the plugs from their sockets—and then uses each strand as a template on which a new strand is synthesized. The result is two identical copies of the chromosome, because every A attracts a T in the newly made strand, every T attracts an A, every C attracts a G, every G attracts a C. These A-T and C-G combinations are the "base-pairs" of DNA.

On Saturday morning, February 28, 1953, in Cambridge, England, James Watson and Francis Crick realized how the sequence of bases of one strand of the DNA double helix specified the sequence of bases of the other strand. They saw for the first time how the plugs and sockets fit together. When they went to lunch that day at their favorite haunt—the Eagle Pub, not far from their lab—they left the other patrons dumbfounded with their announcement that they had learned the secret of life.

Indeed, Watson and Crick's revelation of the double helix as the structure of DNA, and the realization that the sequence of bases on one strand specifies the sequence of bases on the other, is one of the most important scientific discoveries ever made. Immediately, and very clearly, it explained a major mystery: How does one cell give rise to two identical cells? By revealing the double-helical structure of DNA, Watson and Crick answered a question that had confounded people since before the time of Aristotle: How do organisms replicate themselves? They do it by using one strand of the DNA double helix to specify the sequence of bases of the other strand, producing two identical DNA molecules from one. For their discovery, Watson and Crick were awarded the Nobel Prize in 1962.

With that discovery Watson and Crick ushered in the age of molecular biology, which provided a detailed understanding of the nature of the gene. It culminated in the Human Genome Project, an international effort to determine the sequence of base-pairs in the DNA of every one of our chromosomes. That goal was achieved on the fiftieth anniversary of Watson and Crick's discovery.

Here is a portion of the sequence of bases of one strand of a human chromosome:

CCTTCCGTTAAACCAATGGGAAACAAGTCCCTGGGAGTGTCCCGCCCT
GGGGTGAGAACTGCGAACCAATAAAAATTGAAACCTGAGCGGTGGCGC
GGCCAGCTGTGGGTGGAGTCACCCCGCGGACTGGACGGGAACCTGG
CGGGGTCAGGTCCCGTCAAGCAGCCTGGCTCATGGCTGTGTGCGGCC
TGGGGAGCCGTCTTGGCCTGGGGAGCCGTCTTGGCCTGCGCGGGTGC
TTCGGCGCCGCCAGGCTCCTGTATCCCCGTTTCCAGAGCCGCGGCCC
TCAGGGCGTGGAAGACG

This string of three hundred bases is but a tiny fraction of the sequence of one strand of human chromosome 12, which, at over 132 million base-pairs (actually, 132,349,534 base-pairs), is a medium-sized chromosome. The letters of chromosome 12, if written in the size of the letters of this book, would fill fifty-two thousand pages.

Each human cell carries twenty-three pairs of chromosomes, forty-six chromosomes in all: one set of twenty-three came from Mom; the other set of twenty-three came from Dad. Altogether, we have about 6 billion base-pairs of DNA, 3 billion in each of the two sets of chromosomes, in every one of the approximately 100 trillion cells in our bodies. It's a prodigious amount of DNA: laid out end to end, the DNA from one human would reach to the sun and back more than sixty times. These three billion base-pairs constitute the genetic material that is the human genome, the material that directs the production and maintenance of each of us. But don't let these large numbers intimidate you. Remember, as complex as the complete human genome is, and with biologists still at the earliest stages of deciphering the instructions embedded in its sequence of base-pairs, DNA is just a long chemical, and a simple one at that: just strings of A's, C's, G's, and T's.

Asimov may never have taken time out of his writing schedule to exercise or take vacations, but he would always make time for a good meal. At the age of fifty-seven, and perhaps as the direct result of consuming a giant slice of cheesecake, Asimov suffered a heart attack that hospitalized him for three weeks. He continued to experience angina over the next several years, the pain becoming so severe that even walking became a chore. In November 1983, his doctor advised a triple-bypass operation. Given the choice of waiting until after Christmas or having the operation right away, Asimov chose right away. But he worried that might prevent him from attending the annual banquet of the Baker Street Irregulars, his fellow Sherlock Holmes aficionados, which was to be held on January 6. He had prepared a song for the banquet, and although he expected to be there to sing it, he prepared a taped version that he gave to his wife, just in case.

The evening before his operation, Asimov dreamed that he died on the operating table and that consequently his wife had to play the tape for the Baker Street Irregulars, who stood in tears and applauded for, lo, twenty minutes. But Asimov survived, "and my first thought was that now I wouldn't get the kind of applause I would have gotten if I had been dead. 'Oh—[expletive deleted],' I said in disappointment."

Although Asimov's operation was a success, the blood transfusion that he received was contaminated with the human immunodeficiency virus (HIV), because blood was not then routinely tested for its presence. After suffering numerous medical problems in the years after his surgery, Asimov learned in 1990 that he had AIDS. He died in April 1992 from heart and kidney complications, the true cause of death not being revealed until ten years later when his wife published *It's Been a Good Life,* composed of excerpts from his three autobiographies.

Since you're reading a book about genes, and in particular their role in disease, you may well be wondering why the first disease we mention is AIDS. Surely AIDS, which ranks among the most virulent *infectious* diseases that humankind has faced, is not *genetic* in origin, you may be thinking. AIDS is spread by sexual contact, blood transfusions, contaminated needles, and passage of a fetus through the birth canal of an infected mother. But rare is the disease that escapes the influence of our genes. So we can tell you the following quite confidently: If Isaac Asimov had had a mutation in both copies of his *CCR5* gene—a mutation that resulted in the removal of thirty-two base-pairs of DNA—he would not have contracted AIDS. This gene, identified in the 1990s, specifies a protein that sits on the surface of cells of the immune system, looking for a signal that invaders have breached the lines of defense. The HIV virus uses the CCR5 protein as a landing pad, alighting on it before invading the cell. If Asimov had lacked those 32 base-pairs in his *CCR5* gene his immune cells would not have had the HIV landing pad, causing them to be resistant to the virus. Unfortunately, even though the prevalence of this mutation is higher in the Ashkenazi Jewish population to which he belonged than in most other populations, Asimov was not so lucky. As a consequence, the world got many fewer Asimov books than it might have.

How do we find the gene responsible for a trait such as resistance to the AIDS virus? How does a gene specify a protein? What do proteins do? What does it mean to have a mutation in a gene, and why does the prevalence of different mutations vary in populations? Read on, and you'll see that these questions have straightforward answers.

3 Proteins Are the Workhorses of the Cell: Misdiagnosis of a Metabolic Malady

Patricia Stallings had had a tough life. She had spent several years on the skids. Homeless much of the time, she found it difficult to take care of herself, let alone the son she had borne out of wedlock. When accused of child abuse for not adequately caring for the child, she gave him up for adoption.

But by the summer of 1989 Patty's life had turned around. She found a good man in David Stallings, and their marriage gave her the kind of life she could only dream about a few years earlier. With their move into a trim, white frame house in a subdivision overlooking Lake Wauwanoka, not far from St. Louis, the Stallings family—Patty, David, and their newborn son, Ryan, born in April—joined the middle class. "That truly was the happiest time of my life," she later reflected. "Everything was perfect. Everything. A new house, a new baby. I mean, what could be wrong?"

Plenty, Patty would soon discover. One Friday evening early in July 1989, three-month-old Ryan threw up his evening meal. He seemed better the next day, but Sunday morning he again could not keep food in his stomach. When he turned lethargic and his breathing became labored, Patty called St. Louis Children's Hospital and arranged to bring Ryan there. She hurriedly strapped him into the car seat and drove the forty miles north to St. Louis, but in the confusion of city traffic she ended up at Cardinal Glennon Hospital, a few miles short of her intended destination. But it was close enough: being a children's hospital, surely its doctors should know what to do for Ryan, Patty thought.

The physicians ordered the usual workup, and when the lab results came back they were shocked: high levels of ethylene glycol—antifreeze—had been found in Ryan's blood. Because Ryan's symptoms were consistent with ethylene glycol poisoning, the attending physician suspected Ryan had been poisoned. He notified authorities, and Ryan was promptly placed in protective custody.

Patty was distraught. She knew she wouldn't harm her son, and she couldn't imagine that David would, either. Why had he been taken from them? She visited Ryan as often as possible, always under the watchful eye of a social worker, except on September 1. That day Patty was left alone with Ryan for several minutes while she fed him from a bottle.

Three days later Ryan again became ill, exhibiting the same symptoms that had led to his first hospitalization. Lab tests again revealed high levels of ethylene glycol in his blood, and the lab technicians identified a trace of ethylene glycol in the bottle Patty had used to feed Ryan. A second lab confirmed the presence of antifreeze in Ryan's blood, and a search of the Stallings's home turned up a gallon jug of antifreeze. Perhaps with her past in mind, authorities arrested Patty and charged her with poisoning her child. By the time she arrived at the jail, her five-month-old son was barely clinging to life. She was forbidden to see him, and on September 7, 1989, Ryan died. Patty was charged with first-degree murder. The prosecutor said he would seek the death penalty.

While in jail, grieving the loss of her son, Patty realized that she was pregnant again. She was still in jail in February 1990 when she gave birth to her and David's second son, David, Jr., called D.J. D.J. was immediately placed in foster care. Not only was his incarcerated mother prevented from seeing him, but his father, too, was denied contact with his son, even though David Sr. had been charged with no crime and had no criminal record.

A few weeks later D.J. became ill, with symptoms remarkably similar to those that Ryan had exhibited before he died. D.J. was taken to St. Louis Children's Hospital (the one to which Patty had intended to take Ryan), where he was eventually diagnosed with methylmalonic aciduria (MMA), a rare hereditary disease.

People with MMA can only partially break down the nutrients in milk and other foods. In D.J.'s case the problem was due to a missing protein that goes by the name cobalamin adenosyltransferase. This protein is necessary to carry out one of the steps in the digestive process, and without it, D.J. could only partially metabolize the milk he was fed. Consequently, toxic byproducts accumulated in his bloodstream. But because he was correctly diagnosed very early in his life, his diet could be modified before the toxic metabolites took their toll, so D.J. survived.

Could Ryan have died because his personal DNA code also resulted in a nonfunctional version of the same protein? Had toxic metabolic byproducts due to MMA, rather than antifreeze, killed Ryan? Had Patty spent seven months in jail for the "crime" of transmitting to her son a gene that specified a defective protein?

What are proteins? What do they do? Why does the absence of the protein cobalamin adenosyltransferase cause children to become sick? When we say "protein" here, we're not using the term in the generic sense of a constituent of our food, as when we say that meat and eggs and nuts contain a lot of protein whereas bread is mostly carbohydrate, and butter is basically fat. In the context here we are talking about individual proteins, of which there are roughly twenty thousand different varieties encoded by the twenty thousand genes in the human genome. Just as DNA is a chemical, each of those 20,000 proteins is a distinct chemical, in this case composed of carbon, hydrogen, oxygen, nitrogen, and sulfur atoms. (But certain foods we think of as protein-rich contain a lot of a particular type of protein: eggs are rich in a protein called albumin; milk is full of a protein called casein.)

While DNA gets all the glory, proteins do all the heavy lifting. Proteins are the tiny machines that carry out nearly every cellular process, working in conjunction with other constituents of the cell to keep it alive and carry out its functions. The proteins in these machines are like gears and flywheels and valves: they fit together with exquisite precision and act in synchrony to carry out a specific cellular task.

Proteins determine much of what we see when we look at someone. They provide the texture to our hair and skin, and the color to our blood. But most of what they do is done quietly and invisibly. Some proteins function to copy the DNA when a cell divides, others break down nutrients into digestible bits, and other proteins use those bits of nutrients to synthesize new cellular material. Yet other proteins are sentinels that monitor the environment and transmit what they learn about it to the interior of the cell and to neighboring cells. Many proteins are enzymes—like the cobalamin adenosyltransferase that D.J lacked—biological facilitators that speed up chemical reactions, like those that occur when we digest food.

A human cell may make on the order of ten thousand different proteins, but most people are familiar with only a tiny fraction of them. These well-known proteins include insulin, which modulates the amount of sugar in the bloodstream, and hemoglobin, which captures oxygen in the lungs and ferries it through the bloodstream to the tissues. Antibodies are familiar proteins that serve as our border patrol, making the rounds of the body to defend us against potential attackers such as invading bacteria and viruses.

Less well-known proteins are the targets of virtually all drugs. Lipid-lowering drugs known as statins, prescribed to bring down elevated cholesterol levels, inhibit a protein essential for cholesterol synthesis (its name is 3-hydroxy-3-methyl-glutaryl-CoA reductase); the pain relievers ibuprofen and aspirin target a protein involved in inflammation (cyclooxygenase-2); Prozac relieves depression by inhibiting a protein whose job is to regulate the level of a chemical, serotonin, that relays signals between brain cells. AIDS has become a treatable disease because drugs are available to block the activity of two proteins, a protease and a reverse transcriptase, that are necessary for the virus to reproduce.

Proteins are the workhorses of the cell. If we think of our body as a diverse company whose mission is keeping us alive and happy, genes are management; proteins are the labor force.

When disease strikes, the immediate cause is usually the absence of a normal human protein, as was the case with David Stallings Jr., or a detrimental change in a human protein. Cancer results from the uncontrolled division of cells, which can occur either because a protein that normally puts the brakes on cell division is defective, or because a protein whose job is to promote cell division is hyperactive. One form of diabetes is due to the failure to make enough of the protein insulin; another form is due to defects in proteins responsible for detecting insulin. Neurodegenerative diseases such as Alzheimer's, Parkinson's, or ALS (amyotrophic lateral sclerosis, known as Lou Gehrig's disease) are still poorly understood, but it is clear that aberrant proteins play a role in most of them.

Disease can also be caused by the presence of a toxic foreign protein. Cholera, diphtheria, botulism, and anthrax are caused by poisonous proteins that are released from bacteria that have invaded the body.

Proteins, like DNA, are polymers, long chains of a few different types of simple chemicals—in this case, small molecules called amino acids. The protein polymer is more complex than the DNA polymer because it consists of twenty different kinds of molecules—twenty different amino acids—rather than just the four types of bases (A, C, G, and T) of DNA. A few of these amino acids, such as tryptophan, have achieved some notoriety as dietary supplements. Others have been implicated in disease, such as the amino acid phenylalanine, which causes severe problems for people with the inherited disease called phenylketonuria. But most amino acids remain well off people's radar screens.

Proteins range widely in size, from just a few to thousands of amino acids linked together. The typical protein is a chain of three hundred to five hundred of the twenty different amino acids. Just as with DNA, it is the *sequence* of these amino acid subunits—their exact order in the protein—that determines a protein's chemical and physical properties.

Sixty years ago biochemists argued over whether each protein has a single unique sequence of amino acids, or is a collection of different amino acid sequences. The English biochemist Fred Sanger was the scientist who settled this argument by determining the order of amino acids in the protein insulin, confirming that their sequence is unique. For that accomplishment Sanger was awarded the Nobel Prize in Chemistry in 1958. (In 1980 he became the only person to win two Nobel Prizes in Chemistry, one of only four people to win two Nobel Prizes in any field.) Despite his remarkable accomplishments, he is known for his modesty and his quiet, unassuming nature. He preferred "to putter about in the laboratory" rather than to be a high-profile globe-trotting scientist.

Sanger determined the sequence of amino acids in insulin because it is a small protein that could be obtained in large amounts because of its medical importance. But small as the insulin protein is, it still has too many amino acids to sequence straight away, so Sanger first chopped it into smaller pieces. He then applied the elegant methods he and his colleagues had developed for identifying the order of amino acids in small fragments of protein. Having established the sequence of amino acids in the protein fragments, he was able to assemble the sequence of amino acids in the whole protein.

The process Sanger used is conceptually simple. Imagine a string of letters whose sequence (their order) is to be solved. The string is chopped up randomly into smaller pieces, and the sequence of the letters in each piece is determined. For example, the pieces may have the following sequences: M-R, L-P-L, R-L-L, L-W, L-L, P-L, W-M-R, L-W-M, L-L-P (each letter is an abbreviation for one of the twenty amino acids). Knowing that these short sequences all come from the same longer sequence, you can line up the fragments:

L-W
L-W-M
W-M-R
M-R
R-L-L
L-L
L-L-P
L-P-L
P-L

and see that this sequence must be L-W-M-R-L-L-P-L. This is the order of a stretch of amino acids in insulin. You undoubtedly appreciate that the longer the sequence gets, the tougher the problem becomes. Eventually Sanger was able to work out the order of all fifty-one amino acids in insulin, thus earning himself a trip to Stockholm.

How is it that the sequence of amino acids in insulin instructs cells to take up the sugar glucose from the bloodstream, whereas a different sequence of amino acids of hemoglobin causes it to ferry oxygen around the body? Both proteins are composed of the same twenty amino acids; it's the different order in which the amino acids are strung together that determines each protein's distinct properties. Each of the twenty different amino acids has a different chemical structure, so each has a different shape and different physical properties, which determines how they interact with each other.

The order of the amino acid subunits in a protein chain determines which amino acids interact with each other to cause it to fold up into its own unique three-dimensional shape. Like the ridges on a key, the shape of a protein is the main feature that determines its function and how it contributes to constituting a creature and sustaining life. That's because proteins are designed to fit precisely with other constituents of the cell, much as a key fits into a lock (see figure). Some proteins that play a role in copying DNA have shapes that match particular strings of base-pairs in DNA; some proteins have shapes that enable them to wrap around carbo-

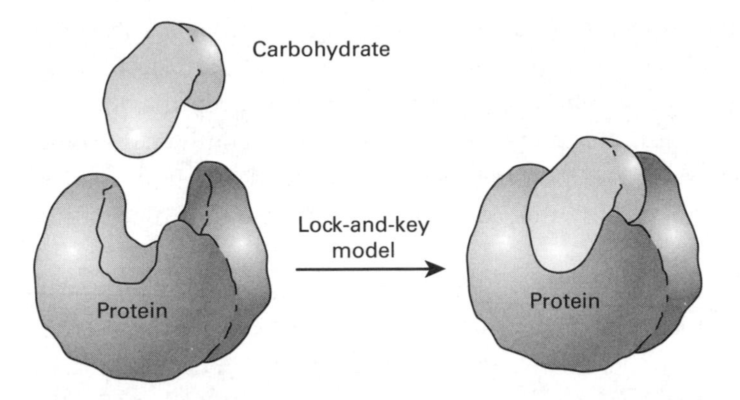

hydrate molecules, which they then cleave into simpler sugars. The protein insulin fits snugly into a pocket of another protein, which then signals that there's too much glucose in the blood. Proteins come in a multitude of shapes and sizes, and those shapes and sizes determine what they can do.

Back in Jefferson County, Missouri, Prosecuting Attorney George B. McElroy III found the evidence against Patty Stallings to be overwhelming. Antifreeze had been found in Ryan's blood on two occasions, by two different diagnostic laboratories, using two different methods of analysis. Those laboratories also found traces of antifreeze in the bottle that Patty used to give Ryan his last meal, and the police found a gallon jug of antifreeze in the Stallings home. Perhaps the most damning evidence against Stallings was the crystals of calcium oxalate found at autopsy in Ryan's brain—a telltale sign of ethylene glycol poisoning.

But since it had been established that D.J. had MMA, a hereditary disease, there was a good chance that Ryan had also had that disease. Could MMA be confused with ethylene glycol poisoning? "Impossible!" said the experts that prosecutor McElroy consulted. They maintained that there was no way MMA could cause high levels of ethylene glycol in the blood. Ryan may have had MMA, they said, but there was no doubt that he had died of antifreeze poisoning. And the Stallings's attorney did not produce any experts to challenge the lab results. The results of the blood tests seemed unimpeachable. It was hard to deny that Ryan Stallings had been poisoned, and Patty was the only person who could have done it. She remained in jail until May 1990, when she was released on bail to await her trial for murder.

How does the sequence of base-pairs in the gene encoding cobalamin adenosyltransferase specify the sequence of amino acids in the cobalamin adenosyltransferase protein? By 1950 it was clear that DNA carries the code for making an organism, and biologists began to focus on how the information specified in the DNA sequence gets converted into proteins. The answer, which was largely worked out by 1965, turned out to be satisfyingly simple.

The DNA sequence does not directly specify a protein sequence. Instead, the process occurs in two steps. First, DNA is copied ("transcribed," in biologists' lingo) into a very similar molecule called RNA (for ribonucleic acid). The RNA is then read ("translated") by a cellular machine to make a protein.

Like DNA, RNA is a polymer consisting of four nucleic acid bases linked together in a long chain. The four kinds of bases in RNA are almost identical to the ones in DNA but not quite: they have one additional oxygen atom. The RNA versions of A, C, G, and T are strung together in the RNA chain in exactly the same order as their counterparts in the DNA that directs production of the RNA copy. The same order of bases is maintained in the RNA because the DNA base-pairing rules (A pairs with T; G pairs with C) also apply to RNA. The protein machine that makes RNA glides down one of the DNA strands "reading" the sequence of bases while synthesizing an RNA copy. When the machine encounters a T, it inserts the RNA version of A in the RNA chain. If the next base it encounters is a G, it inserts the RNA version of C in the growing RNA chain; if an A, it inserts the RNA version of T, and if a C, it inserts the RNA version of G.

RNA differs from DNA in another important way besides the extra oxygen atom: it is single-stranded, a copy of just one of the two strands of DNA. For some genes the RNA-synthesizing machine reads the "Watson" strand of the double helix while making the RNA copy; for other genes it reads the "Crick" strand while making the RNA copy. In either case the product is a single-strand RNA copy of the gene suitable for directing synthesis of a protein.

In the second step of protein production the sequence of bases in the RNA copy of the DNA is translated into a protein sequence by another molecular machine composed of many different proteins. The protein-synthesizing machine reads the RNA bases in groups of three, each succeeding group of three bases in the RNA specifying one of the twenty amino acids at each succeeding position in the protein chain.

Each sequence of three bases, called a base triplet, specifies a particular amino acid. The list of all possible base triplets and the amino acid each of them specifies is the genetic code. It is much like the list of dots and dashes of the Morse code, which specify letters of the alphabet. For example, the triplet AAG specifies—codes for—the amino acid lysine: whenever those three bases appear next to each other in a stretch of RNA, they cause the code-reading machine, traveling down the RNA like a train on its tracks, to insert lysine at the corresponding position of the protein chain. If the next base triplet in the gene is CGA, then the next amino acid added to the growing protein by the code-reading machine is arginine. There are sixty-four possible base triplets (four possible bases at the first position of the triplet times four possible bases at the middle position times four possible bases at the last position of the triplet), but only twenty amino acids, so most of the amino acids are specified by more than one triplet of bases. For example, the amino acid lysine is specified by both the AAA and AAG base triplets. Four of the base triplets have special roles: ATG serves as the signal for the code-reading machine to START making protein, beginning with the amino acid methionine; TAG, TAA, TGA are signals to STOP translating the RNA sequence into protein.

With this knowledge, we can translate into amino acids the beginning of the RNA that gets copied from the DNA sequence of chromosome 12 shown in the last chapter, which encodes the beginning of the cobalamin adenosyltransferase protein. Once the code-reading machine finds the ATG triplet in the RNA template, which tells it to start to synthesize a protein, each succeeding triplet directs the insertion of the next amino acid into the growing protein chain. The beginning of the RNA encoding cobalamin adenosyltransferase directs the synthesis of these nine amino acids in precisely this order at the beginning of the protein: methionine-alanine-valine-cysteine-glycine-leucine-glycine-serine-arginine:

CTGGCGGGGTCAGGTCCCGTCAAGCAGCCTGGCTC ATG GCT GTG TGC GGC CTG GGG AGC CGT ...

methionine alanine valine cysteine glycine leucine glycine serine arginine ...

The code-reading machine marches down the complete RNA template, three bases at a time, using the genetic code to translate each base triplet into an amino acid that gets incorporated into the growing protein chain. Eventually it encounters one of the three triplets that tell it to stop translating the RNA sequence (for this RNA, the code-reading machine would continue for 723 more bases before it encounters a "STOP" triplet, producing a cobalamin adenosyltransferase protein of 250 linked amino acids).

A conceptual framework that may be helpful to understanding the roles of DNA, RNA and protein in the cell has DNA as the wiring diagram for the circuitry of the cell, RNA as the carbon copy of the diagram that gets carried to the fabricators, the genetic code as the legend that reveals what all the squiggly symbols in the wiring diagram mean, and proteins as the switches, batteries, lights, fuses, and other components of the circuits. A mistake in a part of the wiring diagram (a gene) can lead to a defective component (a protein), which can lead to a faulty circuit (disease).

Prosecutor McElroy told the jury: "Don't try to understand why Patricia Stallings poisoned her child by feeding him from a baby bottle laced with antifreeze. The point is she did it. Only she could have done it." After hearing these words, the jury didn't take very long to reach a verdict. A few hours later, on February 1, 1991, the jury foreman, Delmar Fisher, stood before the court and announced the verdict: Patty Stallings was guilty of first-degree murder. A few weeks later Circuit Judge Gary P. Kramer sentenced Patty to life in prison without the possibility of parole. Patty's friends and family sat in the gallery wearing T-shirts bearing the legend "Please help us: Patricia Stallings is innocent."

In fact, help was on the way. Patty's husband, David, had been working hard to get the case more publicity, hoping that someone who was able to help would take an interest in Patty's plight. He managed to get the producers of the TV show *Unsolved Mysteries* interested in the case, and they ran an episode on Patty's predicament in May 1991.

Among those who watched the show was Dr. William Sly, a well-regarded geneticist and pediatrician who was chairman of the Department of Biochemistry at Saint Louis University. As a coauthor of the major textbook on inherited metabolic disorders, Sly well knew how similar are the effects of MMA and ethylene glycol poisoning, and he was very skeptical that Ryan could have suffered from both.

Dr. Sly learned that one of his colleagues, Dr. James Shoemaker, who ran a metabolic testing lab at Saint Louis University, had obtained a small sample of Ryan's blood from one of the labs whose analysis had helped convict Patty. Shoemaker's analysis of the sample also turned up something that looked like ethylene glycol, but only a small amount, nowhere near enough to poison a child. But he saw something else—something that the other two labs had not reported: a large amount of propionic acid.

Shoemaker and Sly knew that propionic acid, which is chemically very similar to ethylene glycol, is a toxic metabolite that accumulates in the blood of people with MMA. Could propionic acid in Ryan's blood have been misidentified as antifreeze? Sly and Shoemaker scrutinized the results from the labs that claimed to have found antifreeze in Ryan's blood, and they were taken aback: the results matched those obtained from a pure sample of propionic acid, and not those of a pure sample of ethylene glycol.

Sly sent a letter to Prosecutor McElroy stating that he was confident Ryan had died from MMA, not from ethylene glycol poisoning. McElroy started to have some misgivings about his case against Patty Stallings, but he was still not convinced of her innocence. What about the ethylene glycol in the bottle Patty used to feed Ryan, and the gallon of antifreeze found in her house? And, most important, how to explain that signature of ethylene glycol poisoning—crystals of calcium oxalate—that the coroner found in Ryan's brain?

The Stallings had fired their first lawyer, and their new lawyer, renowned St. Louis attorney Robert Ritter, asked McElroy: "What would it take to convince you Patty did not poison her son?" The prosecutor said he needed to hear from another expert on metabolic diseases, someone renowned in the field and not associated with the case.

Ritter approached Dr. Piero Rinaldo, a well-respected geneticist on the faculty at Yale University and an expert on inherited metabolic diseases. It didn't take Dr. Rinaldo long to agree with Dr. Sly that both labs that analyzed Ryan's blood misread the results. Their analysis, Rinaldo told *St. Louis Post-Dispatch* reporter Bill Smith, was "totally unacceptable, unbelievable, out of this world. I was astonished. I couldn't believe that somebody would let this go through a criminal trial unchallenged."

Prosecutor McElroy had finally heard enough. On September 19, 1991, two years after Patty was first arrested, after she had mourned the death of her son Ryan, had spent thirteen months in jail, and had never been allowed to spend time with her new son D.J., Patty was absolved of all charges against her. In front of reporters and television cameras, Prosecutor McElroy apologized to Patty and David for what he had put them through: "Unfortunately, we can't undo the suffering that the Stallingses have endured during this entire ordeal. And I apologize to them, both personally, and for the state of Missouri." The subdued smile on Patty's face belied her bittersweet feelings.

What about the traces of antifreeze found in the bottle Patty used to feed Ryan? The bottle had been washed in a dishwasher and filled with infant formula before testing, and the compound identified as ethylene glycol "could have been anything," Rinaldo concluded. "Their approach was: anything that showed up in a certain window in that chromatogram would automatically be labeled ethylene glycol. This is just . . . unacceptable," he said with a sad and disbelieving shake of his head.

And those crystals of calcium oxalate in Ryan's brain? Dr. Rinaldo concluded that they were a result of the ethanol drip used to treat Ryan's presumed ethylene glycol poisoning, an appropriate treatment for that condition, but completely inappropriate for someone with MMA; Dr. Rinaldo suspected it had, in fact, hastened Ryan's death. Two years later the Stallingses received out-of-court settlements for Ryan's wrongful death, from Cardinal Glennon Children's Hospital and from the laboratories that got his diagnosis wrong. The amount of the settlements was not disclosed, but whatever the amount, the money cannot possibly have compensated Patty and David for the loss of their son and their ordeal.

But Patty *was* lucky in one regard: there was only a 1-in-4 chance that D.J. would inherit from both his parents a version of the gene that encodes a defective cobalamin adenosyltransferase. In chapter 7 we'll find out why this is so. If D.J. had been born healthy, there would have been no clue that Ryan suffered from a hereditary disease, and Patty most likely would have remained in jail. Patty beat the odds and was absolved of Ryan's death; D.J. did not beat the odds.

4 All from a Single Cell: How a Fertilized Egg Develops into a Baby

Nine months after a human egg is fertilized, a baby's lungs fill with air and she bawls out her first lusty cry. Just thirty-eight weeks ago she was a single cell, created by the union of one of her father's sperm and one of her mother's eggs. How did one cell give rise, in that short span of time, to an organized mass of human flesh with limbs and lungs in the correct places, with the proper number of fingers and toes, and with eyes and ears and everything else working properly? It seems like a miracle. While it is marvelous, it's not miraculous: biologists have learned the principles of the process that produces a complex organism from a single cell.

This process has intrigued scientists for a long time. An early theory to explain human development, dating back more than two thousands years, is that of preformation. This theory provided a simple answer: we already contain in our bodies very small but fully formed members of the next generation, who merely grow within the mother until they reach the size of a baby able to survive outside the womb. Many scientists thought they saw this tiny person—which they called a homunculus—when they peered at sperm through the first microscopes in the seventeenth century.

This explanation sidestepped the seemingly intractable issue of how complexity unfolds. But there's a big problem with this theory of little people: the homunculus must carry its own sperm that shelter an even smaller version of the person who will be born in the next generation, and that prehomunculus must have in its sperm an even smaller pre-prehomunculus that is to be born two generations hence, and so on ad infinitum for all future generations of humankind. By the nineteenth century embryologists had come to see this fundamental flaw in the preformation theory. The alternative view was that the egg was formless, and, after fertilization, goes through a series of transformations that result in a fully formed individual.

But how? Enter now the fruit fly, *Drosophila melanogaster*. The humble fruit fly seems to appear like magic whenever we leave an open bottle of wine on the table or neglect to toss out a banana peel, calling to mind another discredited theory—spontaneous generation—the idea that life forms can spring from nonliving material. *Drosophila* species are cosmopolitan, having hitchhiked from place to place along trade routes, and spread west in North America with the migration of people and their fruits and vegetables and garbage.

The fruit fly also populates thousands of research laboratories, serving as an ideal subject for the investigation of all sorts of biological phenomena. With its small size (a mere 0.1 inches head to tail), short generation time (just a couple of weeks), large litters (hundreds of eggs per mom), and low feeding and housing costs (quite happy to spend their lives in milk bottles feeding on yeast), *Drosophila* has been a fond object of biologists' attention for more than a century. And it is this fly that has yielded many of the secrets of embryonic development.

That a fly would be key to unlocking the path from egg to adult seemed unlikely in the early part of the twentieth century. Tiny *Drosophila* made its name not in developmental biology but in genetics, while larger animals like the frog and the sea urchin were the darlings of embryologists. For a period of about thirty years, beginning around 1910, researchers in the laboratory of Thomas Hunt Morgan—first at Columbia University, later at the California Institute of Technology—made groundbreaking genetic discoveries using the fly. These included showing that genes lie on chromosomes, uncovering the process by which chromosomes exchange pieces of themselves, and figuring out that sex-linked traits are specified by the X chromosome, discoveries that we will discuss shortly, and that garnered a Nobel Prize for Morgan in 1933.

In the 1940s, following its heyday in Morgan's laboratory, *Drosophila* was eclipsed by even smaller creatures as the objects of geneticists' attention. Taking its place in the new field of molecular biology were the bread mold *Neurospora crassa* and the intestinal bacterium *Escherichia coli* and its viruses. Experiments on these rapidly dividing organisms revealed the nature of the gene, the genetic code, the process of protein production, and the principles of gene function.

Beginning in the 1970s, *Drosophila* began its comeback, led by a young German biologist, Christiane Nüsslein-Volhard, who dazzled developmental biologists with her work showing how a single cell turns into a fully formed organism with trillions of cells. In partnership with a young American biologist, Eric Wieschaus, Nüsslein-Volhard tackled a project so audacious in its concept that another geneticist wondered, "Does she have the whole German army working for her?" But it was just Nüsslein-Volhard and Wieschaus, sitting across from each other at a small table in their lab in Heidelberg, Germany for an entire year isolating mutant flies—ones with changes in their DNA sequence that produce deformed embryos—in the hope that learning what goes wrong in each mutant would reveal how the normal flies do it right.

Nüsslein-Volhard and Wieschaus's mutant flies, first described in 1980 in the international scientific journal *Nature*, were crucial to solving the mystery of development, because they led to the identification of the key proteins that decide each cell's fate by turning particular genes on or off. The two biologists analyzed the flies' cells as an investor might analyze a new company to predict whether it is going to be successful: identify the key executives, find out what critical decisions they are making, and observe how the company responds to their strategic mistakes. Nüsslein-Volhard and Wieschaus were shrewd investors: their acumen won them the 1995 Nobel Prize in Physiology or Medicine.

What was striking about Nüsslein-Volhard's approach was its simplicity: it required only a commercially available chemical to cause mutations in the flies, an ordinary microscope for observing the fly embryos, and standard genetic analysis—all of which were available as far back as 1930. Why did no one think to try this approach in the intervening four decades?

Nüsslein-Volhard had been trained as a biochemist; she wrote her doctoral dissertation on her studies of an RNA-synthesizing enzyme from bacteria. She turned to *Drosophila* because she wanted to apply genetics to the problem of development, and found that she "immediately loved working with flies. They fascinated me, and followed me around in my dreams." As a newcomer to the field of developmental biology, Nüsslein-Volhard was unencumbered by the constraints that limited the thinking of other scientists interested in these problems. "I, compared to other people working in this field, came up with ideas. They were blocked in their minds. Other biologists would say, "'This is not done. We don't do that in our field.' . . . I did things that were completely unconventional."

Nüsslein-Volhard knew that the different types of cells in an organism are different because they deploy (biologists say "express") different sets of genes to produce different kinds of proteins. How did she know that? Isn't it possible that a cell develops into a liver cell rather than a lung cell because it possesses a different set of genes than the lung cell? Might not unspecialized cells of a developing organism lose genetic information as they divide, retaining different sets of genes that determine the type of cell they will eventually become? A cell destined to become a liver cell might retain only those genes that are needed to specify a liver cell; a cell destined to become a lung cell might retain a different set of genes that cause it to become a lung cell. Might it work like that?

This very reasonable idea—that development of one cell type might proceed by loss of genes for all other cell types—was ruled out in the 1960s by the English scientist John Gurdon, who showed that specialized cells possess all the genetic information necessary to specify all the other types of cell in an animal; specialized cells do not lose any genes while assuming their particular identity. Gurdon established this principle with a clever and technically impressive experiment with frogs: he used the genetic information present in a single specialized frog cell to program the development of an entire frog. He was the first person to clone an organism.

Gurdon began by removing the chromosomes from an unfertilized frog egg, literally reaching into the egg with a very thin straw and sucking out its nucleus, the part that contains the DNA wrapped up into chromosomes. He then put into that DNA-denuded egg a nucleus he had similarly extracted from a cell he obtained from a frog's gut. If the specialized gut cell had acquired its identity because it retained only gut cell genes, then its genetic material should not be able to program that egg to develop into a frog, because it would be missing genes necessary for making other types of cells. But Gurdon saw complete, normal frogs develop from some of the eggs he had manipulated. He concluded that specialized cells carry all the genetic information necessary to specify an entire animal.

Because Gurdon did these experiments with frogs, his conclusion was met with some skepticism. Some people said the rules for frog development might be different from the rules for other animals. Frogs, after all, are cold-blooded—very different from us and our warm-blooded cousins. Thirty years later, Ian Wilmut, a Scottish veterinarian, quieted any remaining doubters when he cloned a sheep. Employing Gurdon's methods, Wilmut replaced the nucleus of a sheep egg with a nucleus taken from a cell of an adult sheep's mammary gland. The reprogrammed egg was placed into a ewe who served as a surrogate mom, and five months later Dolly burst into the world, proving that the mammary gland cell carried all the information necessary to create a fully formed, normal lamb. This experiment has since been successfully repeated with many other kinds of animal using several different kinds of specialized cell as the source of the nucleus that programs the egg, proving beyond a reasonable doubt that virtually every one of our cells carries the same complete set of genes.

If all cells in an organism have the same genes, specialized cells must acquire their particular identity by using only some of those genes. A liver cell is what it is because it uses only a subset of its genes, those that provide the proteins that make a liver cell and carry out its tasks. It does not express genes for making brain or bone cells. Lung cells deploy a different set of genes, which give them their unique characteristics; they do not express genes for making skin or spleen cells. And so on for the hundreds, perhaps thousands, of different types of cells in our bodies. So now the key question becomes: How do cells in the developing embryo come to use some genes but not others and thereby become a specific type of cell, eventually leading to the organized mass of tissues we call an organism?

As we discussed in chapters 2 and 3, genes are stretches of DNA that contain the information for making a protein. Genes direct the synthesis of a protein by first being copied into a molecule called RNA, which is very similar to DNA but consists of only one strand of bases rather than the two of the double helix. This first step in the expression of a gene is carried out by a special protein machine in the cell that transcribes the sequence of the DNA into RNA copies that are then translated into proteins, much as medieval monks transcribed sacred texts onto papyrus for translation by their colleagues.

Only some of the genes in each cell are used like this: maybe only ten thousand of the twenty thousand or so genes in each cell get expressed. Each gene has a switch that controls whether it is "on" or "off." If the switch is in the "on" position the gene will spring into action and be transcribed into RNA; if the switch is in the "off" position the gene will remain at rest. The switches of some genes are in the "on" position only in muscle cells, while the switches of other genes are flipped "on" only in nerve cells.

What determines whether a gene's switch is on or off? The decision is made by a class of proteins that we can think of as the executives: their job is to decide whether certain genes are to be on or off. They do this by recognizing and binding to specific DNA sequences near particular genes and regulating their transcription into RNA. Hence their name: transcription factors.

Each transcription factor recognizes one particular short DNA sequence (usually six to twelve base-pairs in length) that is present near the genes it controls. A remarkable property of transcription factors is that they can find their short recognition sequence among the other three billion base-pairs of DNA in the human genome (see figure). They rapidly search through the genome—much the way Google searches through billions of web pages—until they find their sequence, and then they glom on to it.

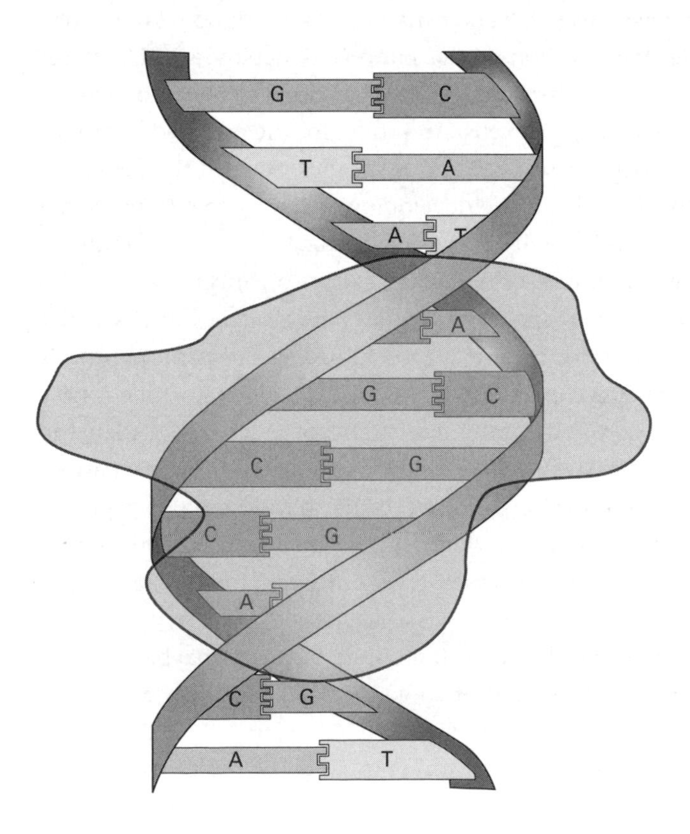

Most genes contain recognition sequences for several transcription factors. The sum of the effect of each transcription factor bound to the gene determines the state of the gene's switch. Some transcription factors act to turn transcription on, others strive to turn it off. The transcription factors are like the transistors that constitute the motherboard of a computer, integrating the input they receive and responding with the coordinated output you see on your screen. This integrated circuitry of transcription factors bound near a gene constitutes the switch that turns the gene on or off.

Actually, these switches are more like rheostats that can be turned up or down, the brightness or dimness of the rheostat's setting being determined by the particular combination of transcription factors that are bound to the gene. Since the human genome encodes about fifteen hundred different transcription factors, the number of different combinations of them is huge, so the rheostats can be set to an almost limitless number of levels. And since the settings of the rheostats on all 20,000 genes determine the identity of a cell, the great diversity of cell types in the human body should no longer be a surprise.

Wise investors know that too many executives often spell doom for a company, so we may wonder why successful organisms such as humans have so many transcription factors. But a complex organism has to make many more decisions than even the largest of companies, and we need all those transcription factors to do that. The factors ask questions about what's going on inside and outside the cell: Are there enough nutrients? What are the cells next door up to? Is there a big demand in the rest of the body for things this cell makes? And many, many other important questions.

The transcription factors learn the answers to these questions, integrate that information, and take action by turning on the genes that are needed (and turning off those that are not needed) by a cell that finds itself in that specific situation at that particular time. The diversity of transcription factors allows many questions about cellular fitness to be asked simultaneously and continuously. The answers to those questions comprise a huge amount of data that the transcription factors process in deciding which genes should be active, and thus which proteins will be present at that specific time in that particular cell.

The decision to turn a gene on or off is like the choice an editor must make whether to run a story about a big fire with a banner headline on page 1 or to go with a more modest mention on an interior page. The editor gets input from several reporters: some at the fire watching a rescue in progress, others at the mayor's press conference hearing what the city's emergency teams are doing, and a few at the hospital listening to the stories of victims. The editor integrates this input and decides the story will run on page 3, but with a large headline. Transcription factors are the cells' editorial staff, collecting information from a press corps of proteins that gather a huge amount of news as they survey the situation.

We can see, then, that cells become different by expressing different sets of genes, which results from each kind of cell having a unique collection of transcription factors. Liver cells have a corps of transcription factors that turn on genes necessary to make a liver cell and turn off genes necessary for making other cell types; lung cells have a different corps of transcription factors that are responsible for turning on the genes lung cells need and for turning off the genes used by other types of cells.

The whole developmental program, from the first division of the fertilized egg to the birth of a fully formed organism consisting of trillions of cells, is largely a diversification of the transcription factor collections in cells as they divide. How does a cell that is destined to contribute to the iris of the eye come to possess *just* the right set of transcription factors to ensure that the genes for making an iris (and *not* genes for making a retina, or lens, or cornea) are expressed at *just* the right levels and at *just* the right time as the embryo develops? In other words, how do different cells come to possess different transcription factors?

Cells assemble their complement of transcription factors by expressing the genes that encode those transcription factors. Each transcription factor is a protein encoded by a different gene, and the subset of those transcription factor genes that a cell expresses determines the collection of transcription factors it contains.

What determines which transcription factor genes get expressed in a cell? The same mechanism that determines the expression of every other gene in the cell: the particular collection of transcription factors it contains.

Oh, oh . . . We seem to have boxed ourselves into a corner: different cells express different genes because they possess different combinations of transcription factors. But they possess different combinations of transcription factors because the genes that encode those transcription factors are acted upon by yet other combinations of transcription factors. A bit circular, isn't it?

The genes that encode transcription factors, like all genes, have rheostats that govern their output, and, as is the case for all genes, those rheostats are set by transcription factors. So the set of transcription factors that a given cell has at any given moment is the result of which particular transcription factor genes were expressed during the course of development of that cell. This logic makes the developing organism seem like a set of nested Russian dolls: to determine why a set of transcription factors came to be present in a liver cell, you have to look at the transcription factors in the cell that gave rise to the liver cell, and to determine why that particular set of transcription factors came to be present in *that* cell you have to look at its precursor cell, and so on, all the way back to the original fertilized egg.

That is precisely what Nüsslein-Volhard set out to do: go all the way back to the first few cells of the embryo and identify the transcription factors they have that make them different, then learn how the cells produced in successive divisions come to possess different combinations of transcription factors that cause them to express unique sets of genes and thus become increasingly specialized.

The genes she discovered that control this process operate by a few general principles. While these principles are simple, the complex process of development is anything but. We'll illustrate the principles that govern development in the fly; human development operates a bit differently, but the fundamental principles are similar.

One principle is that the fly egg, even before it ever sees a sperm, is already subdivided into specialized areas: one end will give rise to the head, the other end to the tail; one part will become the top of the fly, another part the bottom. The basis of this polarity of the egg is chemical gradients in the egg.

Certain proteins in the egg get synthesized by the mother at one end of the egg—say, the end that will become the head—and their levels diminish as they spread outward toward the other end of the egg. If one of these proteins is a transcription factor, it would be most effective

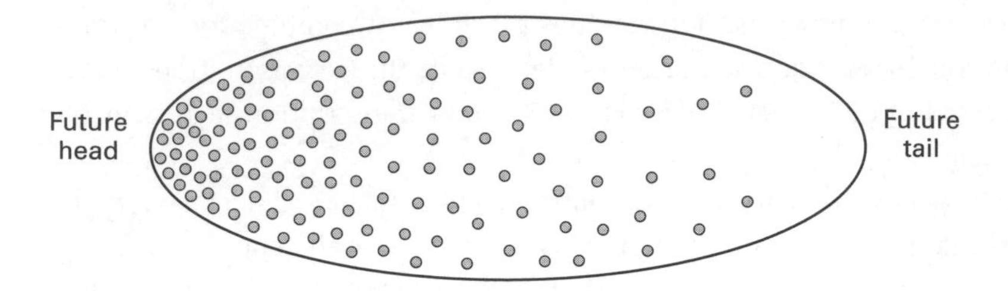

controlling the activity of genes in cells that lie near the end of the embryo destined to become the head, with decreasing effectiveness as its concentration diminishes toward the end of the egg that will form the tail (see figure).

This can be visualized by imagining that you've opened up a can of blue paint in preparation to repaint your kitchen. It sits peacefully in a corner while you gather the brushes and track down the tarp. Just then your teenager zooms in to demonstrate her latest skateboard maneuver, tipping over the can as she glides across the room. The blue paint pours across the floor, a thick puddle in the region nearest the corner where the can stood, thinning out as it spreads across the floor. There is now a gradient of paint that spreads from one end of the kitchen to the other.

A second principle is that different genes respond to different amounts of a transcription factor. One gene might need a high level of a transcription factor to be turned on, a level present only at the region of the egg that will give rise to the head. This may occur because the DNA sequences in the gene that that transcription factor binds to are not very good matches to the sequence it recognizes, so that many copies of the transcription factor are necessary to ensure that some of them recognize and latch on to the partial recognition sequence. Another gene might contain a DNA sequence that is a close match to the sequence recognized by that transcription factor and may therefore require less of the transcription factor to be switched on. As a consequence, that gene will be turned on in cells farther away from the head-forming end of the embryo.

A concentration gradient of a single transcription factor will already define three zones of the fertilized egg: a zone of high concentration at one end of the egg (say, where the head of the fly will form), where the factor turns on genes containing strong and weak recognition sequences

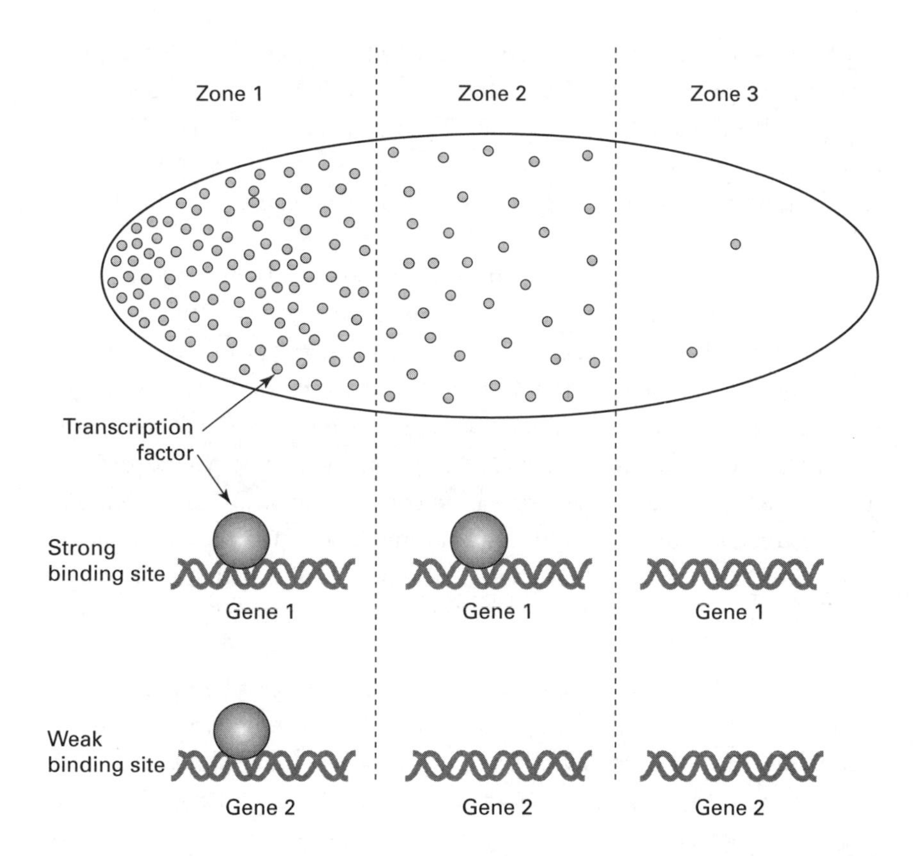

for the transcription factor; a zone of medium concentration near the middle of the egg, where it turns on only genes that have close matches to the recognition sequence (strong binding sites for the transcription factor); and a zone of low concentration near the opposite end of the egg (where the tail of the fly will form), where there is not enough of the transcription factor to turn on either kind of gene (see figure).

A third principle is that cells talk to one another, and these conversations influence which genes get expressed, much as conversations in the hall of a high school influence who is going to the prom with whom. Neighboring cells communicate with each other through proteins they display on their cell surfaces, which act like molecular feelers, or antennae. When these antennae make contact with a neighboring cell, or detect molecules given off by neighboring cells, they send signals into the cell that affect the function of certain transcription factors that result in changes in gene expression.

Among the genes whose expression is affected by these signals are those that encode transcription factors. Since cells in different parts of the developing embryo get different cues from their neighbors, their antennae generate different signals, and thus different cells come to express different sets of transcription factor genes, which eventually cause them to express different genes, which determine the fate of the cell.

These kinds of intercellular conversations are constantly going on in the developing embryo. It's a veritable cacophony. Let's listen in: "Hi neighbor! I've decided to become a cell of the iris, but I can't form the iris all by myself so I'd like you to join me. Hey, you over there! Listen up and get with the program! I'm sending you a signal, so pay attention. And after you receive it make sure you pass it on to your neighbors. We're going to need some of them to become cells for a cornea." By means of these intercellular conversations cells continually refine the set of transcription factor genes they express, ultimately causing them to express the specific set of genes that results in their taking on very specific functions.

Why was it that Christiane Nüsslein-Volhard, rather than some other biologist, had the idea to seek the *Drosophila* mutants whose analysis would reveal these principles? Evelyn Fox Keller points out that as a German scientist, Nüsslein-Volhard was less affected by the gene-centric view of biology typified by the Americans, and was more willing to consider how the other components of the cell participated in the process. Furthermore, as a molecular biologist she was impatient, unlike many developmental biologists; she was accustomed to getting quick results from her experiments. Most critically, she had the imagination to come up with novel ideas.

If everything goes right with the gradients of transcription factors, with the combinatorial interplay of proteins sitting on the DNA, with the intercellular conversations and negotiations, and with the many other things that go into the developmental process, then a complete organism is eventually born, with limbs and lungs in the right places and with the proper number of fingers and toes and with irises and eyelids that work.

Most of the time it does go right—remarkably so, given the complexity of the process. But things can go wrong. A very large percentage of human pregnancies—perhaps 30 to 50 percent—spontaneously abort before the pregnancy is detected because something goes very wrong soon after fertilization of the egg. Fifteen to 20 percent of known pregnancies also result in miscarriages, most of them probably due to mistakes in the developmental program. And the parents of one out of every twenty-eight babies get the distressing news that their child has a birth defect. But when one considers everything that must go right with the process for a healthy child to be born, it's remarkable that we're here at all, let alone with all of our organs and limbs in their proper places and working correctly. And all from a single cell!

5 When the Gene Is the Cure: Immunodeficiency and Gene Therapy

David Phillip Vetter could not live like this any longer. His doctors knew it; his parents knew it; he knew it. They all agreed he had to risk the bone-marrow transplant. Without it he would have to continue living in the bubble—his sterile isolation chamber—waiting for a cure to be developed for his affliction. Because David suffered from Severe Combined Immunodeficiency (SCID), he had no immune system to fight off even the most timid of invaders. He had already waited for twelve years, and still no cure for his condition was in sight. On October 21, 1983, he received some of his sister's bone marrow. It didn't take. Worse, it gave him cancer. He died February 22, 1984, 15 days after walking out of his bubble for the first time.

The first son of Carol Ann and David Vetter Jr. also began life with no immune system, and died of a massive infection six months after birth. His personal DNA code included an X chromosome, inherited from his mother, that carried a defective copy of the gene called *IL2RG*, which provides the instructions to make a protein required for the immune system to develop properly. Because there was a mutation—a change in the DNA sequence—in the *IL2RG* gene David Joseph inherited from his mother, the gene directed the production of a nonfunctional protein. Without the IL2RG protein, David Joseph's thymus, a small organ near the lungs where immature white blood cells from the bone marrow bivouac before going into battle, could not send off white cells to fight infections.

After their experience with their first son, Carol Ann and David Jr. understood that if their next child were a son, he would also have a 50 percent chance of being born with no immune system. A son has only the single X chromosome he inherits from his mother, his other sex chromosome being the Y chromosome he inherits from his father. So if one of the genes on the X chromosome were defective, he would suffer the consequences (see figure). A daughter would be safe, because even if Carol Ann gave her the X chromosome with the defective gene on it, her father, David Jr., would provide another X chromosome carrying a good version of the gene. (In chapter 7 we discuss in more detail why one good gene may be all you need.)

Diseases like SCID that are due to a defective gene on the X chromosome are passed to boys only from their carrier mothers, who have a good version of the gene on their other X chromosome. Males with the mutant gene on the X die of the immune disease before they are old enough to reproduce and pass the flawed X chromosome on to their daughters.

But the Vetters were told that even if their next son were unlucky and drew the defective chromosome, he would not necessarily be doomed: The doctors thought they could cure his disease, either with a bone-marrow transplant from his sister or with a cure they thought was just around the corner. They had on their team Dr. Raphael Wilson, an expert in germ-free environments, who would build and maintain the sterile isolation chamber—the "bubble"—that would protect their infant son from the germs that had killed his brother.

A few years before, Wilson had reported stunning success in Germany with a sterile isolator he built for twins with immunodeficiency: after a short time in the bubble their immune systems suddenly, and inexplicably, came to life, and the twins were taken out of the isolator. So the Vetters' doctors were optimistic. Wilson "just swept us along with his enthusiasm. He had the confidence to say, 'We can do this. We can do this,'" said Dr. Mary Ann South, one of the members of the medical team, in a documentary film about the child who became known as the Bubble Boy.

Carol Ann and David Jr. were eager for another child. Although they had a healthy girl, they wanted a boy to carry on the Vetter family name. "Children were very essential to our hope and to our dream of the future. We wanted to have children right away; we wanted to have as many as God would send us," Carol Ann explained in the TV documentary. So they had another child. Happily, one of their dreams came true: it was a boy. Sadly, their other dream did not: the boy did not inherit his mother's functional *IL2RG* gene. Instead, like his older brother, he inherited her X chromosome that carried the defective gene. He was whisked into the isolator within seconds of his birth, and that's where he stayed for twelve years, until it became obvious that no cure was imminent.

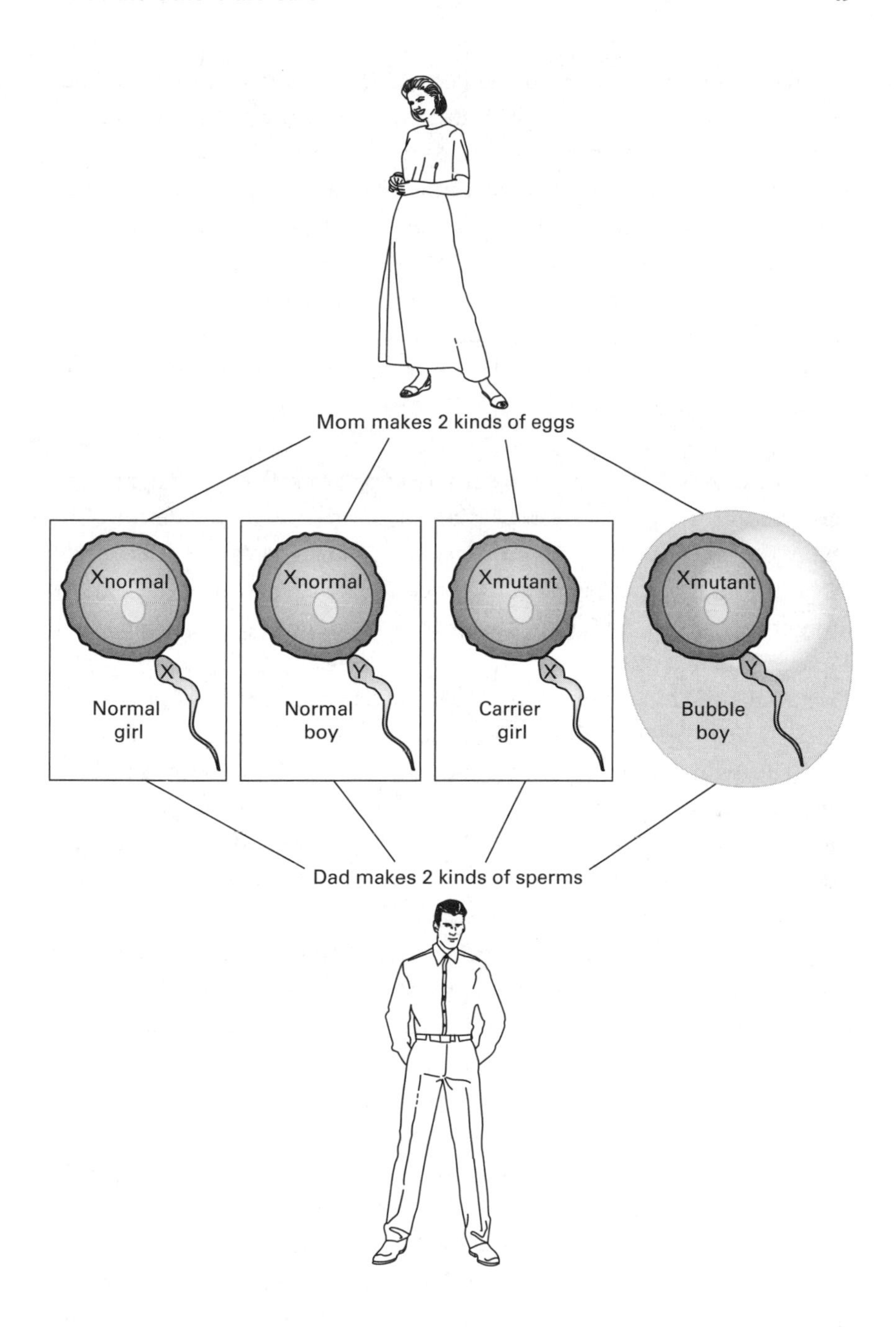
Mom makes 2 kinds of eggs
Xnormal
X
Normal girl
Xnormal
Y
Normal boy
Xmutant
X
Carrier girl
Xmutant
Y
Bubble boy
Dad makes 2 kinds of sperms

David Phillip Vetter, the Bubble Boy, lived a celebrated life that stimulated a hit song by Paul Simon, feature films starring John Travolta and Jake Gyllenhaal, and an episode on *Seinfeld*. Celebrated, but tragic. The journalist Steve McVicker described in a 1997 article in the *Houston Post* how David responded when his friend the psychologist Mary Murphy asked him why he was so angry: "Why am I so angry all the time? Whatever I do depends on what somebody else decides I do. Why school? Why did you make me learn to read? What good will it do? I won't ever be able to do anything anyway. So why? You tell me why!" Murphy had no answer for David Phillip.

Had David Phillip Vetter been born twenty years later he might have chosen to wait a little longer, because in the year 2000 a cure for SCID finally became available. Not a perfect cure, but there can be little doubt David would have jumped at the chance to try it. The cure comes in the form of the good *IL2RG* gene—the gene whose lone copy on David's lone X chromosome didn't work.

If a functional copy of the gene can be delivered to the bone-marrow cells of a SCID patient, those cells begin to make white blood cells competent to fight infections, giving the patient something he wasn't born with: a functional immune system. Treating disease with genes—gene therapy—is the brass ring that David and his parents and his doctors were waiting for. It cured the disease for thirteen boys in France and England. But gene therapy came too late for David.

What occurred after David's death in 1984 that made gene therapy a viable treatment for his disease by 2002? Lots. The human genome was mapped, making it possible by 1987 to identify the region of the X chromosome that carries the *IL2RG* gene. These maps, as we'll discuss in chapter 12, show the positions of genes along a chromosome, just as roadmaps show the positions of cities along a highway. By 1993 scientists had isolated the *IL2RG* gene, using methods for isolating genes developed in the 1970s. In the 1990s scientists devised methods to deliver genes to human cells, so by 1999 they could deliver the *IL2RG* gene to the bone-marrow cells of five children with SCID. By 2002 it was clear that most of these children were cured: four have a nearly normal immune system and are enjoying what David longed for: a life outside the bubble.

How, exactly, was all of this done? Once a gene is located on a chromosome, how is it purified and isolated in the test tube? The principle is simple: the chromosomes are fragmented into small pieces of DNA, and the piece containing a particular gene is fished out of the mixture and copied millions of times, in a process called cloning. It's like making copies of an animal, as was done to clone the sheep Dolly, but in this case multiple identical copies (clones) of the gene are made from a pure template. Each copy is a clone of the original gene that provided the template.

Gene cloning is not much different from what you do when you include a passage from Shakespeare in your wedding announcement. You open your massive compendium of the bard's plays and search through it, page by page, until you find the specific sequence of letters you desire: "Doubt that the stars are fire; Doubt that the sun doth move; Doubt truth to be a liar; But never doubt I love." You extract that passage, insert it into your announcement card, and make many copies of the card to send to friends and family. You have cloned a passage from *Hamlet*, act II, scene ii.

Because of remarkable technical advances of the 1970s, it's now almost as easy to find and copy genes as it is to find and copy passages from a book. The first step is to chop all the chromosomes into small pieces, which can be accomplished by adding to the chromosomes enzymes that cut DNA. Those enzymes don't cut the DNA just anywhere. They recognize specific short sequences of bases in DNA and cut wherever those sequences occur, producing a discrete set of fragments (see figure).

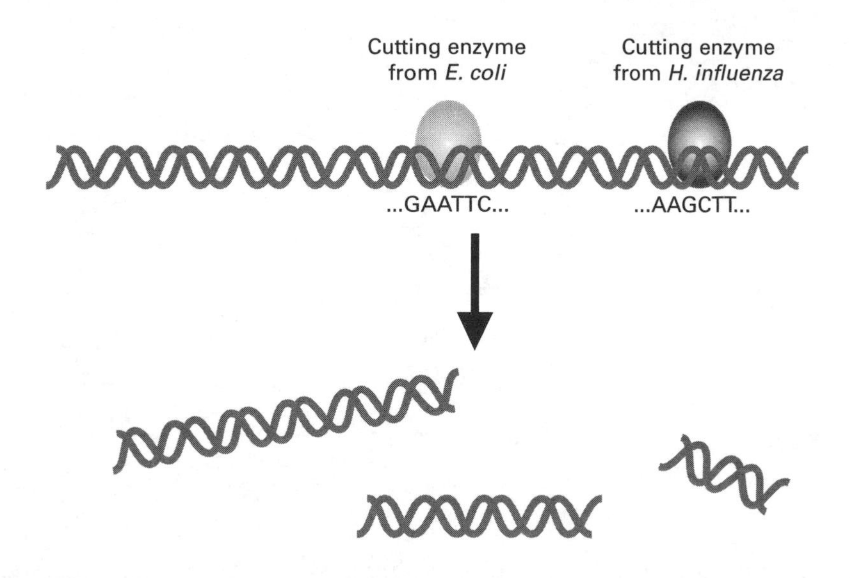

These enzymes are obtained from bacteria, where they provide defense against invaders such as viruses that infect bacteria by cutting up the viral DNA to prevent the viruses from commandeering the bacterial cell. Each species of bacteria has a unique set of these enzymes, and each enzyme cuts DNA at a different sequence of DNA base-pairs. For example, an enzyme in the common gut bacterium *Escherichia coli* cuts DNA wherever it finds the base sequence GAATTC on one of the strands; an enzyme from the bacterium *Hemophilus influenzae*, which causes pneumonia, cuts DNA wherever it finds the base sequence AAGCTT. Over three thousand of these enzymes have been characterized, and most are readily available, allowing gene hunters to divvy up the chromosomes into bite-sized pieces—basically dividing the book that is our genome into paragraphs.

Having cut up the genome, we need to separate all the fragments—there are millions of them—much as we separate the pieces of a jigsaw puzzle before starting to piece them back together. The fragments are first inserted, one by one, into a minichromosome, a small circular piece of DNA derived from a bacterial chromosome, using another kind of enzyme that joins two pieces of DNA, as an ironworker constructing a skyscraper joins two steel beams (see figure). All cells have such a DNA-joining enzyme because they constantly need to unite pieces of DNA to repair the damage that DNA continually incurs.

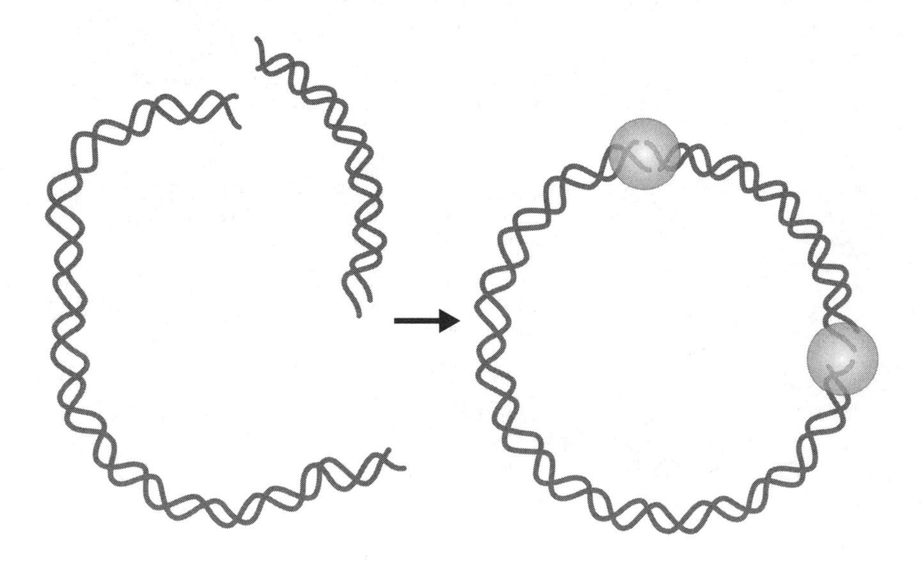

The minichromosomes are slipped back into bacteria, which act like little copying machines to make copies of the minichromosomes that carry a piece of a human chromosome. Each bacterial cell divides to produce a colony of billions of cells, each carrying a minichromosome containing a particular piece of the human genome. Among those millions of identical-looking colonies of bacteria, we need to identify the one that carries the minichromosome with the fragment that includes the gene we want. How can we do that?

We know the DNA sequence of the *IL2RG* gene because the base sequence—the precise order of A's, C's, G's, and T's—of the human genome has been determined, so the process now is simple: the base-pairing principle worked out by Watson and Crick means that any DNA strand will match up to its complementary partner strand through the specific plugs and sockets on each base. For instance, the sequence AGGCTAAC will match up to TCCGATTG because A pairs with T, and C pairs with G. So we can design and have synthesized a piece of DNA with a sequence of bases that matches up to some sequence in the *IL2RG* gene. Such synthetic DNA molecules can be ordered from providers of scientific supplies for not much more than you'd pay for a T-shirt. Because that piece of DNA will stick to the *IL2RG* gene, it can be used as a "probe" for the gene. We will attach to that probe a molecule that can be easily detected, like a compound that is fluorescent. When we spread some of this fluorescent DNA on top of the millions of bacterial colonies, it will find its mate only in the rare colony of bacterial cells that carries the *IL2RG* gene on its minichromosome, and the probe will literally "light up" that cell. We can then recover the cells of that glowing bacterial colony and retrieve the minichromosome from them, with not much more difficulty than we retrieve copies from the output tray of a Xerox machine.

The same enzymes that enable gene isolation spawned the biotechnology industry, today a worldwide enterprise that generates over $50 billion in yearly revenue. This new industry has provided such drugs as erythropoeitin (Epo) and granulocyte colony stimulating factor (G-CSF) for stimulating the growth of blood cells in patients who have undergone chemotherapy for cancer, and antibodies such as the one that fights breast cancer by inhibiting the Her2 protein, and insulin for diabetics, and Etanercept for treating disorders of the immune system such as arthritis and psoriasis, and the blood-clotting Factor VIII for hemophiliacs, and many more.

In addition to enzymes that split apart and splice together DNA molecules, there are enzymes that can duplicate DNA sequences to make more copies of them, enzymes that can change one sequence to another, and enzymes that carry out some of the many steps required to manufacture pharmaceuticals. None of these enzymes was sought by biologists to create an industry. Rather, they were discovered—quite fortuitously—in scientists' quest to understand how bacteria fight infection, or how they replicate their DNA, or how they synthesize their proteins, and many other seemingly esoteric questions. Clearly, basic research is a good value.

Soon after he was born in Buckinghamshire, England in August 2003, Alexander Locke was diagnosed with the same disease that killed David Joseph and David Phillip Vetter. Alexander's parents, Carol and Colin Locke, like the Vetters, had no idea their firstborn son was at risk of having SCID. "We realised Alexander had a problem when his tummy button inexplicably failed to heal after birth, despite repeated courses of antibiotics. At four months, he developed a severe viral respiratory infection. He spent his first Christmas in hospital, attached to oxygen lines and antibiotic drips," Colin told Andrea Kon, a reporter for England's *Daily Telegraph*. "He had inherited his defective X-gene from me," said his mother, "and it was hard to accept that it was my 'fault.' I had no idea I carried a 'bad' X-gene."

Alexander was put in an isolator in London's Great Ormond Street Hospital for Children. It was more comfortable than David Phillip Vetter's bubble because the technology had improved in the intervening twenty years: Alexander had an entire room to romp around in. Alexander was protected by an airlock through which all his visitors had to pass in order to have the air around them cleaned and filtered. He was to live in the isolator while he waited for his doctors to identify a perfectly matched bone-marrow donor, something for which David Phillip Vetter waited for in vain for twelve years.

But Alexander spent only eight months in the isolator. Drs. Adrian Thrasher and Bobby Gaspar at Great Ormond Street Hospital were getting ready to test an experimental gene therapy for treatment of SCID, and when Alexander's bone-marrow transplant fell through (because the nearly perfectly matched donor carried a virus that would almost certainly have killed him), Alexander entered the gene therapy trial, along with four other boys with SCID.

Drs. Thrasher and Gaspar had developed a vehicle to deliver to Alexander's bone-marrow cells a good version of his defective gene. The vehicle was a virus that infects human cells. Viruses are ideal for this job because they are basically tiny Trojan horses that carry DNA within their protein coat. The viral DNA contains genes, just as our DNA does, and these genes code for the viral proteins that make up the protective coat and that make copies of the viral DNA. Although the viral DNA is minuscule compared to ours—most viruses have just a handful of genes and often only a few thousand DNA base-pairs—scientists have found places in this DNA where other genes, human genes, can be inserted.

The virus enters a cell and removes its coat, thereby delivering its DNA inside the cell. If it's a normal virus, the cargo is the viral chromosome with its genes that encode proteins to commandeer the cell's machinery to make more virus. Some viruses are aggressive, making many copies of themselves and killing their host cells in the process, releasing more viruses that go off to infect and kill other cells. Other viruses are relatively benign, incorporating their own DNA into a human chromosome while allowing the cell to live, lying in wait to check out the situation before deciding to make more virus. But if some of the viral genes are removed from its chromosome, the virus is disabled: it can deliver its DNA into cells, but that incomplete viral chromosome cannot take over the cell or produce more virus.

Drs. Thrasher and Gaspar spliced the *IL2RG* gene into the chromosome of a disabled virus that infects human cells. They made many copies of the engineered viral chromosome, packaged them into viral coats, and added the viruses to a test tube containing Alexander's bone-marrow cells. The viruses latched on to the marrow cells and quietly slipped into them, taking off their viral coats as they went in.

The viral DNA made its way to the cell's nucleus where it pasted itself along with the *IL2RG* gene into one of the human chromosomes. As the cells grew and divided they passed the viral DNA along with the good *IL2RG* gene on to other bone-marrow cells. The engineered cells were returned to Alexander's bloodstream, where they found their way back to his bone marrow.

Many processes have to go right for gene therapy to work, and scientists are still a long way from having a failure-proof procedure. For gene therapy to work, biologists need to deliver the viruses carrying the therapeutic genes to the appropriate cells, where the good gene can do its job. Discoveries of how cells specialize in certain tasks have led to improvements in this cell targeting. Even if the good gene gets to the right cells, it must get turned on at the right time and at the right level to provide cells with the right amount of the protein they are missing, when it is needed. Furthermore, expression of that gene must persist for long periods of time, so understanding how transcription factors and other proteins determine whether a gene is on or off is invaluable.

Drs. Thrasher and Gaspar waited to see if the engineered bone-marrow cells would give rise to the white blood cells Alexander needed to fight infections. "Alexander was allowed home for the first time in May, aged eight months. It was more complicated than having a newborn," Carol told Andrea Kon. "Every tube and piece of equipment needed sterilising. We had to use a spreadsheet to keep track of his medical regime. He had lost the ability to suck during the first days in intensive care and was being fed through a gastronasal tube. We were administering four drugs four times a day through the tube."

Despite their best efforts, Alexander contracted infections and had to be rushed back to Great Ormond Street Hospital. "The second time doctors fought a nine-hour battle for his life; we were so lucky. It's astounding he didn't suffer any long-term brain damage," Colin exclaimed.

Alexander's new bone-marrow cells grew and began to spawn competent white blood cells. He was allowed to venture out of his airlocked room for longer and longer periods of time. "He loved mixing with other children, although he had never played with a child until six months ago," his mother said. "We feared that his life as a 'bubble baby' might have left developmental or physical scars, but he's caught up in every way. The only treatment he needs now is a prophylactic course of antibiotics once a fortnight. Soon he'll go to primary school. That would have been unthinkable two years ago." The cure that David Phillip Vetter was waiting for finally arrived. It came in the form of a gene.

But good as it seems, the cure is far from perfect. The virus that delivers the *IL2RG* gene to bone-marrow cells of SCID sufferers can also deliver something deadly: cancer. Four of eight boys in France who were cured of SCID by gene therapy traded it for another disease: leukemia. One has since died from it. The cancer occurred because the piece of viral DNA that carries the *IL2RG* gene, which usually inserts itself benignly into apparently nonfunctional regions of the genome far from any critical genes, landed in these boys' genomes near a gene encoding a protein that accelerates cell growth. The viral DNA caused this gene to be turned on, and the protein it made set those cells on the path to cancer.

The doctors had no way of knowing in advance that the virus would land near this gene, but as soon as they learned that it had, they stopped doing gene therapy with the virus while they searched for a way to prevent it from happening again. Is the prospect of a cure for their disease worth the risk of leukemia for these boys? We suspect that David Phillip Vetter would have said it is.

It is not surprising that the only real successes of gene therapy since David Phillip Vetter's death in 1984 have been with disorders, like SCID, that affect cells that doctors can easily get their hands on. Surgeons are good at harvesting bone-marrow cells from patients, and scientists are adept at growing them and modifying them in the laboratory. And it is easy to get the engineered cells to the place they are needed, because when they are reintroduced into the bloodstream they find their way back to the bone marrow, like salmon returning to their home river to spawn.

Other attempts at gene therapy have not been so successful. Dr. Ronald G. Crystal at the U.S. National Institutes of Health had a great idea for delivering to patients a good copy of the gene that is defective in people with cystic fibrosis. This gene encodes a protein that sits in the membrane of lung cells and allows salt to pass in and out. If the protein is defective and the salt balance is upset, then the layer of mucus that keeps germs out of the lungs becomes thick, providing an attractive breeding ground for infectious bacteria. The inflammation of the airway that results makes breathing difficult. Cystic fibrosis is a disease that usually leads to an early death.

Crystal reasoned that he could deliver the gene to patients' lung cells simply by having them inhale disabled viruses that carry the cystic fibrosis gene. The viruses would be sucked into the lungs, where they would attach to cells and inject the functional gene. He used a relatively harmless virus that naturally infects the lungs and gives people mild cold symptoms. While the idea seems terrific, it didn't work because the lung cells are protected by an armor of mucus and cilia, little hairs that sweep away foreign particles that enter the lungs, which ended up blocking access of the viruses to the cells.

One gene therapy failure in 1999 was especially tragic. Jesse Gelsinger, at eighteen, suffered from a metabolic disorder caused by the lack of the ornithine transcarbamylase (OTC) enzyme, which is needed to prevent a toxic product of metabolism, ammonia, from accumulating in the blood. The accumulation can lead to brain damage, coma, even death. Jesse had a mild form of the disease, which he was able to keep under control with medication (he took thirty-two pills a day) and a low-protein diet. He knew gene therapy was unlikely to help him, but he was eager to try it because it promised to help those with more severe forms of the disease. Jesse told Sheryl Gay Stolberg, a reporter for the *New York Times*, "What's the worst that can happen to me? I die, and it's for the babies."

At 10:30 a.m. on Monday, September 13, 1999, a large dose of a virus carrying the *OTC* gene was injected into a vein that emptied into Jesse's liver. The plan was that it would deliver the gene to his liver cells, which would then make the enzyme he needed. We'll never know if that happened, because Jesse died four days later, the victim of a massive reaction of his immune system to the virus. His death cast a pall over gene therapy for several years.

Given its few successes and its several failures and tragedies, gene therapy has yet to live up to its much-ballyhooed potential. There are still enormous challenges in getting functional versions of genes into the right cells and, once there, getting them to produce an appropriate amount of the needed proteins for long periods of time.

And gene therapy brings enormous ethical questions. If we can change someone's personal DNA code by replacing a defective *IL2RG* or *OTC* gene in order to cure a disease, we could probably change a gene to make a child taller, or stronger, or have a lower level of cholesterol, or concentrate for longer periods of time. What are the limits on human characteristics that are permissible to alter? And what about making changes to the DNA code that would be passed down to future generations? Although there is general agreement that this type of gene therapy should never be attempted, human history suggests we must remain vigilant. Scientists are nothing if not persistent and ingenious, and they have no lack of alternative strategies to someday bring gene therapy into standard practice. But the expectant public that has learned of the potential of gene therapy to relieve suffering from diseases, as well as scientists themselves, must be patient. For David Phillip Vetter the wait would have been twenty years for a possible cure; for other patients with other diseases, the wait will be longer. But the cures *will* come. Of that we are confident.

6 When Cells Are the Cure: Diabetes and Stem Cells

Disease is egalitarian: it strikes the privileged as well as the downtrodden, the youthful as well as the aged. Few children born in 1907 had the advantages in life of Elizabeth Evans Hughes. The daughter of Charles Evans Hughes, governor of New York, who later served as secretary of state and as a distinguished Chief Justice of the Supreme Court, Elizabeth was a bright and vivacious young girl. But by the summer of 1922 fourteen-year-old Elizabeth stood five feet tall yet weighed only forty-five pounds. She was so emaciated, her muscles so wasted, that she could barely walk. Elizabeth was dying of diabetes.

At the time of Elizabeth's diagnosis physicians had come to appreciate that diabetics could not properly metabolize their food and consequently were subject to weight loss, weakness, gangrene, infections, and other symptoms that often caused death within a year of diagnosis. Elizabeth's doctor, Frederick Allen, advocated a starvation diet for diabetes, which could hold off symptoms of the disease but resulted in a patient's wasting away. Diabetics faced a cruel choice: death by diabetes or death by starvation. Elizabeth's parents chose for her the latter, and the severe diet she was put on, at times supplying her with a meager three hundred calories a day, kept Elizabeth alive for three years after her symptoms became apparent.

Elizabeth had type 1 diabetes, also known as juvenile diabetes because it primarily strikes the young. Much more common is type 2 diabetes, also known as adult-onset diabetes because it appears in adulthood. Both types are characterized by the inability of cells to take up the sugar glucose from the blood following the digestion of food. Consequently, glucose accumulates to high levels in the blood, spills over into the urine, and is lost. The increased glucose load requires a larger-than-normal volume of urine, leading to terrible thirst. Not getting enough nutrients, a diabetic becomes weak and tired, and is constantly hungry. Add to these problems the fact that glucose is actually a poison when it is present in the blood at higher-than-normal levels, and we can see why Elizabeth's disease was so devastating.

Glucose is a sugar that is the primary energy source for the cells of our body. Much as gasoline is burned by an automobile to produce the energy that propels it, glucose is "burned" to produce the energy that cells need to carry out their work. Some glucose we obtain directly—for example, from the sugar we put in our coffee or from the coating on our breakfast cereal—but most of it is obtained from the more complex food we eat as it is broken down by specialized cells in our gut.

The glucose that is released quickly finds its way to our bloodstream, where it flows to all cells in the body. Cells are prompted to take up glucose from the blood by the hormone insulin, a protein produced by the pancreas and released into the bloodstream after a meal to signal that glucose is now plentiful and needs to be disposed of.

Type 1 diabetes occurs when the immune system runs amok and attacks and kills the insulin-producing cells of the pancreas, so that little or no insulin is made, preventing other cells in the body from receiving the signal that glucose is available. Type 2 diabetes is due to an inability of cells to recognize the insulin signal, so they don't realize that they should take up the glucose that is in the bloodstream. As the glucose levels in the blood continue to rise, the pancreas must produce more insulin to keep up with the ever more urgent requirement to dispose of the extra glucose. The pancreas gets increasingly fatigued, and eventually gives up, leading to insulin-dependent diabetes, with its dire consequences.

By 1920, researchers knew that diabetes was due to a failure of pancreatic function: in diabetics, the pancreas still pumped digestive juices into the intestine, but it did not produce the secretion that went into the bloodstream to regulate metabolism. This secretion held the key to a possible therapy for diabetes, but its identity and nature were as yet unknown. At the University of Toronto, a young surgeon named Frederick Banting had a novel, albeit somewhat misguided, idea about how to obtain this internal secretion in an intact form, protected from the destructive effect of the pancreas's digestive juices.

Banting planned to test his idea on man's best friend. Beloved for their companionship, loyalty, and fun-loving nature, dogs have sometimes had to serve not as pets but as subjects of studies designed to cure serious diseases. Large enough for surgery, with a physiology that is a good match for that of humans, they are used when other animals such as mice won't do. Banting's plan was to tie off the ducts of the dogs' pancreases that empty the digestive juices into the gut, so as to prevent those juices from destroying what the pancreas secreted to manage blood glucose levels.

Banting believed that the cells making the digestive juices would wither away, while those that make the secretion that regulates glucose levels would be spared. He would then prepare an extract from these pancreases. At the same time, he would make other dogs diabetic by removing their pancreases, and then try to cure their disease by injecting them with the pancreas extract. Banting, together with his student, Charles Best, began injecting the diabetic dogs with pancreatic extracts, monitoring the dogs' blood glucose levels to determine whether the extract was effective.

It worked just as they had hoped: during a frenzied period of experimentation beginning in the summer of 1921, Banting and Best saw that the glucose levels in the blood of their diabetic dogs were lowered after they were injected with the pancreatic extracts. Extracts of cow pancreases (available in abundance from the nearby slaughterhouse) worked too. The success with the cows' pancreatic extracts indicated that the digestive ducts of the dogs' pancreases did not have to be tied off in complicated surgery. As Banting and Best later learned, there was never any danger that the destructive enzymes made by the pancreas would destroy the insulin in the secretion because they are produced in inactive, harmless forms that become active and destroy any proteins they encounter only after they are far from the pancreas.

But Banting never realized that his initial stroke of insight that ultimately led to the discovery of insulin was misguided. He went to his grave believing that a crucial contribution to the discovery of insulin was his idea to cripple the pancreas so that it couldn't secrete the digestive enzymes that (he thought) destroyed the prize he was after. As happens with many scientific breakthroughs, Banting got the right result for the wrong reason. Some scientists see their mistakes; others don't. Scientists can be as stubborn as anyone, and they often hold on to an idea long after others have shown it to be far from reality. Yet even wrong ideas, if well implemented, can lead to great good.

Banting and Best were soon joined by the biochemist J. B. Collip, who worked out a procedure to purify the precious material, which they named "insulin." The name is based on the Latin root for the word "island," as the substance is derived from a special cluster of cells that form an island floating in the sea of the cells that make up the rest of the pancreas. At the end of 1921, Banting traveled to New Haven, Connecticut, to announce the exciting results of his experiments in a session for diabetes experts attending the annual meeting of the American Physiological Society. By May 1922, the Toronto group had engaged Eli Lilly and Company to produce insulin at the company's facilities in Indianapolis. Working with impressive speed, Lilly sent its first shipment of insulin to Toronto in July.

Banting examined the barely alive Elizabeth Hughes on August 16, 1922, and immediately started her on insulin. By the end of a week of treatment, she was put on a diet of 1,220 calories a day, and a week after that she was eating a normal diet of over 2000 calories. Even more remarkable to her, Elizabeth could again eat foods such as white bread, corn, and macaroni and cheese, all forbidden to her under her treatment by Dr. Allen. Descriptions of the first diabetics receiving insulin in the 1920s read much like those of AIDS patients in the first days of highly active retroviral drug treatment: a near miraculous recovery from the worst symptoms, a regain of weight back to their old levels, and a return to normal daily life.

Type 1 diabetes, accounting for 5 to 10 percent of diabetes cases, generally occurs in young people of normal weight. The antibodies that attack and destroy the insulin-producing cells of the pancreas may be triggered by a viral infection. Symptoms of the disease develop quickly, and treatment invariably requires regular injections of insulin.

The much more prevalent type 2 diabetes (about 90 percent of cases) has until recently mostly struck older people, the great majority of whom are overweight. The onset of the type 2 form is gradual, often requiring several years to become full-blown. Treatment includes weight loss, increased physical activity, blood glucose testing, and medication, either oral drugs or injected insulin.

The total incidence of diabetes is staggering: more than 7 percent of the American population has the disease! As we increasingly give in to the drumbeat of the fast-food marketers to consume more high-fat, high-calorie foods, more and more of us, including our children, are learning that we have diabetes. We may like the convenience and low price of a quick meal, but what marketers avoid telling us is that eating too many burgers and fries and the like carries a heavy price. As affluent lifestyles, with their attendant high-fat, high-calorie diets, spread to countries such as India, whose traditional diet was lean, diabetes has become a worldwide epidemic.

As with most diseases, our personal DNA code plays a large role in our risk of contracting diabetes and in the course the disease takes if we get it. The genetics of diabetes is especially complex. The disease clearly runs in families: siblings of diabetics have a much higher risk of developing the disease than does the general population, and the identical twin of a diabetic—who shares her identical personal DNA code—is affected much more often than is a fraternal twin, who has a different personal DNA code.

The versions of genes that make up our DNA code influence other factors besides the initial risk of getting diabetes: how glucose is metabolized, what level of body mass is attained, and how the body responds to insulin are also highly heritable. It is not surprising, then, that some ethnic groups—who, as we'll discuss later, are more apt to share particular versions of their genes than are unrelated people—have relatively high rates of the disease: type 1 is especially frequent in people of European descent; type 2 is more prevalent in African Americans, Hispanic Americans, Native Americans, certain Asian Americans, and Pacific Islanders.

More than half of the genetic risk for type 1 diabetes is determined by a region of chromosome 6 that carries over one hundred genes. Many of these genes encode proteins that function in the immune system and might be responsible for its inappropriate attack on the insulin-producing cells that is the cause of the disease. These proteins sit on the outside surface of cells, acting as sentinels that continually examine proteins in the body and identify them as friend or foe. If the proteins are recognized as one's own, they are allowed to go on their way, but if they seem to come from an invader such as a virus or bacterium, they set off an alarm that triggers an immune response designed to eliminate the threat. Sometimes the alarm is so loud that other proteins—such as those in islet cells of the pancreas—are caught in the frantic response, and type 1 diabetes ensues.

In the case of type 2 diabetes, multiple genes contributing to the disease—or in many cases just regions of chromosomes where such genes may lie—have been identified, but their effects are usually small, and in some cases still uncertain. It seems likely that a large number of genes, influenced by such outside determinants as the type of diet and the amount of exercise a person gets, contribute to the onset of the disease and determine how it will progress. So far eleven genes have been conclusively identified as playing a role in type 2 diabetes, and more are sure to be forthcoming.

People with type 1 diabetes can be treated with insulin, but they can hardly be called cured. Diabetics must carefully monitor the level of glucose in their blood and inject themselves with insulin several times a day. But no matter how diligent they are in their surveillance, they must still closely control their diet and maintain their exercise regimen.

Even those who are able to maintain good control of their blood-glucose levels have difficulty keeping them constant; they experience episodes of extremely low blood sugar, which can result in loss of consciousness, and bouts of high blood sugar, which can eventually lead to blindness, cardiovascular disease, and other serious health problems. As a result of these complications, the life span of diabetics may be shortened by as much as one third.

Type 2 diabetes can often be controlled by diet and oral pharmaceuticals, but almost 20 percent of people with this form of the disease eventually require insulin therapy. Type 2 diabetics also develop an array of medical problems, most notably heart disease, depending on how well they are able to control their blood-sugar levels. Banting and Best's breakthrough saved the lives of Elizabeth Hughes and millions of others, but scientists are still searching for the new approaches that are desperately needed to treat and ultimately cure the disease.

Since the insulin-producing cells of the pancreas are destroyed in both types of diabetes, the disease might be cured if these cells could be replaced. Transplantation of pancreatic cells or of the entire pancreas from cadavers has been attempted, but there is a severe shortage of donors for this procedure.

Even if a donor can be found, a formidable roadblock is presented by the recipient's immune system, which recognizes the new cells as foreign and attacks them, and which therefore must be suppressed. But treatments for immune suppression often lead to other, quite unwelcome, complications. And even if the immune system can be fooled and the complications avoided, the refurbished pancreas provides insulin for only a limited amount of time, most recipients needing insulin injections again within a few years of receiving the transplant. Unfortunately, pancreatic cell transplantation is probably not a cure for this disease.

A wholly new avenue that shows promise is regenerative medicine: engineering a special type of cell—stem cells—to treat disease. Stem-cell therapy has been widely discussed in both scientific and public forums. If its promise is realized it could turn our aging bodies into a patchwork quilt of all kinds of different types of engineered cells. What are stem cells and how can we engineer them to provide material for organ or tissue replacement? What is the potential for this type of therapy in the treatment of human diseases?

As we saw in our discussion of human development in chapter 4, a fertilized egg goes on to produce a baby containing hundreds of different cell types by a continuing process of cell specialization. Once each cell becomes specialized—as a blood cell or a lung cell or a brain cell or a muscle cell—it is committed to that fate: there is no going back to the unspecialized, jack-of-all-trades state.

Stem cells, by contrast, have the extraordinary property of remaining unspecialized. But they can be coaxed to become any of several different specialized cell types. A stem cell can divide into two cells that remain as unspecialized stem cells over long periods of time, but under certain conditions a stem cell divides to produce a daughter cell that goes off into the world to make something of herself, embarking on a path toward specialization that eventually leads to life as a skin cell or a heart cell or any one of myriad other types of cell (see figure).

Stem cells exist in the embryo as well as the adult, but not all stem cells are created equal. Embryonic stem cells have the potential to give rise to every possible cell type and tissue: liver and lung, iris and intestine, skin and skeleton, and, of course, the pancreas and its insulin-producing cells.

As adults we also retain many kinds of stem cells in our bodies after our physical development is complete, but those are of limited potential. Adult stem cells have already started down the path of specialization, and as a consequence are restricted to becoming a narrow range of cell types. Adult stem cells in the bone marrow continuously produce new blood cells to

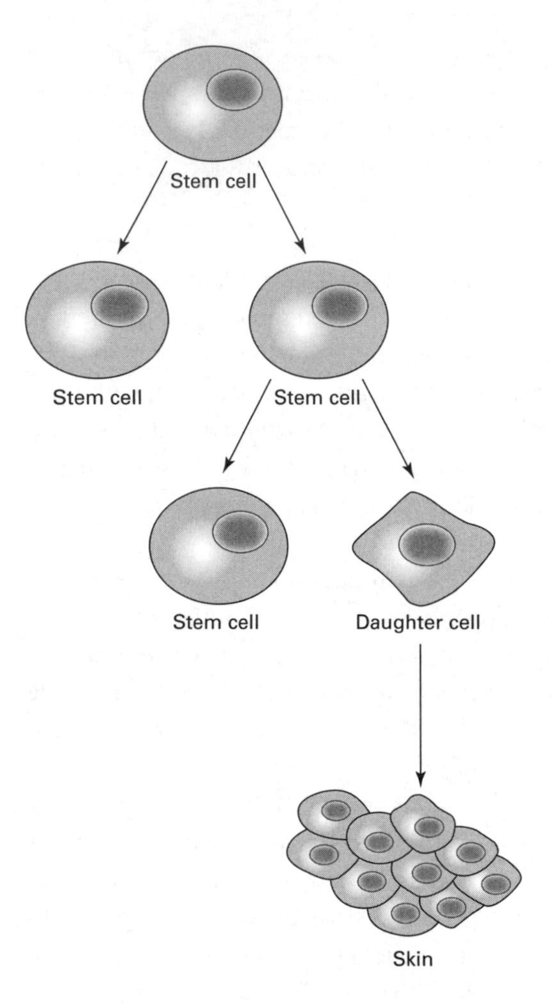

carry oxygen and fight infections; adult stem cells in the liver replenish the ones we damage on New Year's Eve; adult stem cells in the intestine and skin replace those that are continually sloughed off; adult stem cells in muscle provide replacements to the cells that are torn when we lift weights. We constantly lose cells for all kinds of reasons, and adult stem cells come to the rescue and provide replacements.

Even though certain adult stem cells can replenish certain organs throughout our lives, they cannot take up the slack when other types of cells are lost. Stem cells in the hair follicles of the skin give rise to hair and to epidermis, but not to brain cells; stem cells in the lining of the digestive tract give rise to several kinds of gut cells, but not to bone cells. The existence of stem cells for the pancreas is controversial, and even if they do exist it is uncertain whether they can be isolated and used to repopulate the pancreas.

Far more versatile are embryonic stem cells. These cells come from embryos only four or five days old—one source is left-over embryos from in vitro fertilization procedures that have been donated for research purposes. At that stage of development the embryo has only about one hundred to two hundred cells, and those cells have not yet begun their journey along the paths to specialization, so they are able to give rise to all of the cell types found in the adult. These stem cells could potentially be an unlimited source of insulin-producing cells.

But they are not available yet. A number of hurdles must be cleared before embryonic stem cells can be used to cure diabetes. First, new embryonic stem-cell lines must be obtained, because almost all the existing ones are flawed. Many seem to have limited potential, being able to give rise to only certain types of cells. Worse, many of the stem-cell lines that exist today are contaminated, precluding their use in humans. They were grown on a layer of mouse cells, to provide factors necessary for stem cell growth. Those growth factors have now been identified and can be provided in pure form. Biologists have learned much about the care and handling of embryonic stem cells in the last several years, promising healthier, more versatile, and more consistent stem-cell lines in the future.

Second, embryonic stem cells need to be coaxed into specializing as insulin-producing pancreatic cells, and the persuasion has to be gentle, done in a way that mimics the process that takes place in the developing embryo. As embryonic cells specialize to become pancreatic islet cells, different genes are expressed in a carefully choreographed process directed by a few key proteins—the transcription factors we learned about in chapter 4—that recognize and bind to sequences of DNA in the genes necessary to produce pancreatic cells and control the expression of those genes. Activation of such genes eventually leads to production of the proteins that turn a cell into one that produces insulin.

Third, once conditions are established for coaxing stem cells to specialize in insulin production, the process needs to be made easy and reliable. Huge numbers of cells will be required for therapy, so growing them in small plastic dishes as is done today will not be up to the task; cells will need to be grown in large vats to meet the need. Furthermore, the treatments to direct cell specialization must become highly efficient, because cells that resist the nudge and remain unspecialized have the potential to grow into tumors when implanted into a recipient.

Finally, the problem of rejection by the recipient's immune system must be overcome, so that individuals being treated do not destroy the insulin-secreting cells they receive. A possible solution is to establish banks of embryonic stem-cell lines from different people. A doctor treating a diabetic would withdraw from the bank a cell line that is compatible with the recipient. An alternative would be to genetically engineer changes in the cell surface proteins (the ones monitored by the immune system) of embryonic stem cells so they are not recognized as foreigners by the patient's immune system.

In addition to the daunting scientific issues, challenging ethical issues surround human embryonic stem-cell research. Fertilized human eggs must be destroyed to produce new stem-cell lines. For those who believe that a fertilized egg has the same standing as a human being, its destruction poses a moral dilemma. For those who believe that an embryo has the full rights of a living human, the use of embryos in research may not be acceptable.

Many others—and we count ourselves among them—believe that scientists have a duty to prevent or alleviate human suffering, with the utmost importance placed on saving the lives of people already among us. An embryo consisting of one hundred to two hundred cells—about the size of the period at the end of this sentence—has no brain or central nervous system and no other organs associated with a person. This lack of sentience is an argument that this tiny embryo should not have the same status and rights as a living person. Of course, the embryo must be given serious respect, and strict controls must be in place to limit its use to research that can benefit humanity. But we should not abandon the tremendous potential of embryonic stem cells for curing disease.

We think reasonable people should be able to agree on guidelines and regulations to govern the use of human embryos. We do not see why the more than 400,000 embryos that currently rest in freezers and are destined to be destroyed or remain frozen indefinitely cannot be donated for research that has the potential to alleviate human suffering.

In our view, embryonic stem cells have tremendous promise, and scientists should be encouraged to pursue this important line of research. In addition to diabetes, a host of other devastating diseases and disabilities—Alzheimer's disease, Parkinson's disease, heart disease, arthritis, spinal cord injuries, burns, and vision and hearing loss, to name just a few—are potentially treatable by regenerative medicine.

A promising alternative to embryonic stem cells has recently burst onto the scene. Scientists have discovered that turning on just a few genes in adult cells that are committed to a specialized function can "reprogram" them to return to an earlier, unspecialized stem-cell-like state. These genes encode transcription factors that act as master switches able to change the fate of cells. The reprogrammed cells, known as induced pluripotent stem cells, or iPS cells, behave much like embryonic stem cells. Because iPS cells are derived from adult cells and thus can be obtained without the destruction of fertilized eggs, they may provide a noncontroversial approach to regenerative medicine. But the arguments about the use of embryonic stem cells are not going to abate soon: if biologists are to learn how closely iPS cells mimic embryonic stem cells, they must be able to produce and study both types of cells. For now, at least, stem cells derived from human embryos should not be abandoned in our quest for cures.

In 1930, after Elizabeth Evans Hughes's miraculous recovery from near death after treatment with some of the first doses of insulin, she married William T. Gossett, a lawyer for Ford Motor Company, and they had three children. In 1980, when the writer Michael Bliss wrote to her husband to ask how Elizabeth had died, he received a response from Elizabeth Hughes Gossett herself. She was in fine health, physically and intellectually, and she met with Bliss to reminisce about her nightmarish youth with diabetes. She died in 1981 at the age of seventy-three, having survived the disease for sixty years, and after taking more than forty thousand injections of insulin.

We look forward to the time when a young person newly diagnosed with diabetes will receive just one injection: of insulin-producing cells derived from embryonic stem cells or iPS cells, and thereafter will enjoy a healthy life without ever having to poke herself with a needle.

II The Inheritance of the Gene

7 When One Gene Is Enough: The Enzyme Missing in an Inherited Disease

The particular genetic endowment we receive from our parents or bequeath to our children can be a powerful motivator in our lives. For those confronted with a devastating inherited disease, the deep-seated questions—why did this disease occur? how did it occur?—can be a potent force driving the motivation to do something about it. That force sometimes leads to undreamed-of accomplishments. It led one woman to worldwide renown.

In 1914, an American woman who had been raised in China by her missionary parents returned to Chinkiang after attending college in Virginia, to care for her ill mother. There she fell in love with and married an agricultural economist from Cornell University who had come to China to teach farming methods. In 1920, at the age of twenty-seven, she gave birth to a beautiful daughter, whom she named Carol. "Mine was a pretty baby, unusually so," she would write much later. "Her features were clear, her eyes, even then, it seemed to me, were wise and calm."

But while Carol's body began to grow, her mind failed to develop. "I think I was the last to perceive that something was wrong," her mother wrote. When the child still was not speaking by the age of three, the woman was comforted by Chinese friends who told her that speech comes at different ages, even though they themselves suspected a deeper problem.

When Carol was nearly four, her mother heard a lecture about retarded mental development in preschool children given by a pediatrician visiting from the United States, and learned the danger signs to watch for: slowness to walk, slowness to talk, and incessant restlessness. The doctor examined Carol and added other concerns: a short attention span, blank eyes, and lack of responsiveness. The woman resolved to take her daughter back to the United States to find out what was wrong with her. She later wrote:

> Then began that long journey which parents of such children know so well. I have talked with many of them since and it is always the same. Driven by the conviction that there must be someone who can cure, we take our children over the surface of the whole earth, seeking the one who can heal. We spend all the money we have and we borrow until there is no one else to lend. We go to doctors good and bad, to anyone, for only a wisp of hope. . . . We crossed the sea and we went everywhere, to child clinics, to gland specialists, to psychologists. I know now that it was all no use. There was no hope from the first—there never had been any.

The end of the journey came at the Mayo Clinic in Rochester, Minnesota. A German doctor, blunt and honest, not meaning to be cruel but unwilling to delude her, delivered the grim news: "'I tell you, madame, the child can never be normal. Do not deceive yourself. You will wear out your life and beggar your family unless you give up hope and face the truth. . . . This child will be a burden on you all your life. . . . Find a place where she can be happy and leave her there and live your own life.'"

With no hope left, the woman endured unbearable sorrow. Her lone consolation was the realization that Carol herself had no knowledge of her deficiencies and was destined to spend her whole life as if in childhood. The mother's sense of hopelessness was intense, and she often thought how much better it would be if Carol died. But she also came to perceive how coping with her daughter was sapping her own life. "For the despair into which I had sunk when I realized that nothing could be done for the child and that she would live on and on had become a morass into which I could easily have sunk into uselessness. Despair so profound and absorbing poisons the whole system and destroys thought and energy."

The woman became aware that she must find an institution for Carol in the United States, one with appropriate companions and future security. She and her husband had become estranged, for he wanted nothing to do with spending money on private care for their daughter, and she realized that financing long-term care would be difficult. "I had found out enough to know that the sort of place I wanted my child to live in would cost money that I did not have. There was no one to pay for this except myself. I must devise means to do what I wanted to do for my child." She became a writer.

Why do you resemble your parents more than you resemble any of the other six and a half billion people on this planet? Why do siblings bear an uncanny resemblance to each other? Why is a woman whose mother and aunt get breast cancer before they are fifty concerned that she may be at high risk to get this disease? Why was Carol born with a metabolic disorder that leads to profound mental retardation? It's simple: our personal DNA code is made up of specific versions of the roughly twenty thousand genes we get from our parents.

Genes specify proteins, which build the intricate apparatuses that constitute our cells, which make up our tissues and organs, our blood, nerves, skin, and hair. You resemble your parents and your siblings because your personal DNA code is composed of pieces of their personal DNA codes: you share with them particular versions of each gene, causing you and them to be composed of very similar proteins. Since those versions of proteins compose and operate you and your parents and your brothers and sisters, you all have similar traits. But you are certainly not identical to your parents or to your brothers and sisters, because your cells possess a unique combination of their personal DNA codes and thus a unique combination of the varieties of their proteins.

How do we explain the patterns of inheritance we observe? What were the chances that you inherited Uncle Bert's large nose, or your mother's sunny disposition, or the colon cancer that has struck so many people in your family? What were the chances that Carol would inherit a metabolic disorder from her parents?

The principles of heredity were worked out in the 1860s by Gregor Mendel, a monk living in Brno, Moravia, which is now part of the Czech Republic. Mendel deduced these principles by examining the characteristics of pea plants—their height, the color of their pods and seeds, and several other such easily observed features—noting what happened to these characteristics in the offspring of pea plants that he planted in his monastery's garden. The principles that Mendel discovered in his work on peas—there are basically only two of them—are straightforward and easy to understand and apply to all organisms, including humans. Because these principles are universal, with nearly all organisms transmitting their genes to their offspring in the same way, we'll discuss them in terms of human traits.

The first principle is that our genes come in pairs because our chromosomes come in pairs. We have two copies of each of 22 chromosomes, plus two copies of the sex chromosomes (two X chromosomes for females, one X chromosome and one Y chromosome for males), for a total of 46 chromosomes. Our chromosomes came from our mother and father, each of whom likewise has 46 chromosomes, but we didn't end up with 92, and our children won't have 184, because we get just half of each parent's chromosomes: sperm and egg cells have only half the number of chromosomes as the rest of the cells of our bodies.

This reduction in chromosome number that occurs when sperm and egg cells are formed is not a random selection of any 23 of the 46 chromosomes. Rather, one of each chromosome pair is taken. Thus, the egg and the sperm cells resulting from this process of reduction have a single copy of each of the 23 chromosomes. When a sperm cell unites with an egg cell, a complete complement of 46 chromosomes is restored in the fertilized egg. If the fertilized egg has 22 pairs plus two X chromosomes, one from Mom and one from Dad, it will become a female. If it has 22 pairs plus an X chromosome from Mom and a Y chromosome from Dad, it will become a male.

A fortunate consequence of having two copies of each chromosome is that we have two copies of every gene (except for genes on the X and Y chromosomes in males)—think of it as a backup if a certain gene is defective. For nearly all of these genes, one copy is sufficient: it doesn't matter if the other copy carries a mutation and therefore produces a defective protein, because the good copy provides a normal, functional protein. Thus, we can sleep soundly, safe in the knowledge that most mutations can't hurt us because we have a backup copy of every gene.

The second principle of heredity is that genes are apportioned to sperm and egg cells independently. That is, a particular gene in a man's sperm cell is equally likely to have come from his mom as from his dad; the same is true for any particular gene that ends up in a woman's egg cell. How this independence occurs—even for genes that reside on the same chromosome—will be considered in chapter 12.

Around the same time that the American mother in China was coping with her daughter's retarded development, another young woman, Borgny Egeland of Oslo, Norway, gave birth to two children, a girl in 1927 and a boy in 1930. Both children failed to develop normally: Liz, at age six, was able to say only a few words; Dag, at four, was unable to talk at all. Borgny and her husband, Harry, visited numerous specialists, but none was able to explain the cause of the children's condition. The parents also noticed that both Liz and Dag produced urine with a strange musty odor, and wondered whether this smell was associated with their developmental defects.

Eventually the Egelands went to see Asbjørn Følling, an Oslo physician with an interest in metabolic disorders. Følling examined the children and, like the other doctors, could not make a diagnosis. But when he conducted a routine test of their urine for the presence of ketones—chemicals that in elevated concentration can be a symptom of many illnesses, including diabetes—Følling made a remarkable discovery. The test he used involves the addition of ferric chloride, an iron-containing compound, to the urine. If no ketones are present, the urine turns brown; if ketones are present, it turns purple. But the urine from the Egeland children did something Følling had never seen: it turned dark green when ferric chloride was added to it, indicating the presence of an unusual chemical.

Følling asked Mrs. Egeland to bring him urine samples every other day for two months. Eventually she brought him a total of twenty liters, a heroic feat considering that Dag was not toilet trained.

Følling purified the substance in the urine that reacted with the ferric chloride and identified it as phenylpyruvic acid. To determine whether other children who were mentally impaired also produced this substance, Følling tested more than four hundred institutionalized patients in Norway. He found eight more with phenylpyruvic acid in their urine, including two other sibling pairs. He also noted other traits in these patients, including fair skin, eczema, broad shoulders, and a stooping carriage.

Følling later identified thirty-four more patients from twenty-two families, and by 1945 he was able to report the genetic basis of this disease, which was given the name phenylketonuria, or PKU. The disease is due to the failure of these children's metabolism to break down surplus phenylalanine, one of the twenty amino acids found in proteins. The result is an increased amount of phenylalanine in the blood, which eventually spills over into the urine as phenylpyruvic acid. The excess phenylalanine leads to severe problems in the developing central nervous system, for reasons that are not well understood even today, resulting in diminished intellectual abilities in affected children.

The specific defect that is the cause of PKU is a mutation—a DNA sequence change—in a gene that encodes a protein known as phenylalanine hydroxylase, or PAH. The defective *PAH* gene directs the production of a defective PAH protein that cannot do its job, which is to serve as an enzyme that converts phenylalanine to another amino acid, tyrosine.

Long before the gene for this enzyme was identified, babies who failed to produce PAH could be identified by a simple test for phenylalanine, first by measuring it in their urine and then later by measuring it in their blood by means of a test that could be done even before infants left the hospital. These advances led to the adoption by the 1950s and 1960s of routine screening of newborns, which became mandatory in many countries.

Children who test positive are put on a phenylalanine-restricted diet, which prevents the severe symptoms of PKU. Nonetheless, because the diet prohibits milk products, meat, egg, beans, lentils, and many other foods, and because some of the food that can be eaten is a bit unpalatable, it is difficult to follow. And even patients who adhere to the diet have somewhat elevated levels of phenylalanine, and often score lower on intelligence tests and have some emotional and behavioral problems. Despite the imperfection of the treatment, discovering the basis of the disease and developing a simple test for PKU (and subsequently for many other metabolic disorders) must nevertheless be considered one of the triumphs of modern medicine.

In the Egeland family lurked a mutation in the gene we'll call *PAH* on chromosome 12. Both copies of the gene in Liz and Dag must have carried this mutation because they didn't appear to have a good backup gene and their cells could produce no functional phenylalanine hydroxylase protein, hence their developmental problems. Their parents, Borgny and Harry Egeland, were free of the disease, but in their personal DNA codes both must have carried a mutation in one of their two copies of the *PAH* gene, and each parent passed on a defective copy of the gene to their children. Borgny and Harry Egeland are "carriers" of the mutation: they show no sign of the disease but carry a mutation in one of their two *PAH* genes.

Both Egeland parents had one normal copy of the gene (PAH^{+}) to direct production of a functional phenylalanine hydroxylase protein, and one copy of the gene with a mutation that directs production of a non-functional PAH protein. We'll denote the gene with the mutation as pah^{-}. The parents did not have the disease because all of their cells possessed functional PAH protein. The amount of good PAH protein that comes from only a single good copy of the gene was enough enzyme for them to thrive.

The effect of the normal PAH^+ gene dominates over the effect of the pah^- gene with the mutation that directs production of a defective PAH protein. That is, the normal PAH^+ gene is "dominant" over the mutant pah^- gene; the pah^- mutation is said to be "recessive" because its effect (loss of PAH protein function leading to abnormal development) recedes into the background in the presence of a normal PAH^+ gene that encodes a functional PAH enzyme. (Throughout this book, dominant genes will be written in uppercase letters and recessive genes in lowercase.)

Only when both copies of the defective pah^- gene are inherited, one from Mom and one from Dad, as happened to Liz and Dag Egeland, is the effect of this recessive mutation manifested with such severe results. Half of their mother's egg cells received the chromosome 12 with the defective pah^- gene, as did half of their father's sperm. The consequence is that half of the time the egg that is fertilized and develops into a child is one that carries a defective pah^- gene from the mother. Half of those fertilizations will be by a sperm carrying a defective pah^- gene from the father; thus, half of one half, or one fourth, of the fertilized eggs will have defective versions of *both* copies of the gene.

We can see this clearly by listing all of the possible combinations of the *PAH* genes the children of Borgny and Harry Egeland could have received:

From Mom	From Dad	Result
PAH^+	PAH^+	Healthy
PAH^+	pah^-	Healthy (carrier)
pah^-	PAH^+	Healthy (carrier)
pah^-	pah^-	Suffers from phenylketonuria

Although statistically only one in four children of two carriers will get the disease, any given family may be lucky or unlucky. The Egelands were unlucky because, by the luck of the draw, both their children inherited two copies of the defective pah^- gene (the fourth combination above). The statistical probability was that three quarters of the Egeland offspring would have been healthy. Fortunate carriers of the pah^- mutation will produce no children with the disease.

If Liz or Dag had been put on a phenylalanine-restricted diet, it's likely that they would have led nearly normal lives and they themselves would have married and had children. Since Liz's personal DNA code included two defective copies of the *PAH* gene, if she had had children with someone having two normal copies of the *PAH* gene, all of her children would have been carriers:

From Liz	From Partner	Result
pah$^-$	*PAH*$^+$	Healthy (carrier)

If Liz's partner was a carrier (unlikely, because the mutation is rare in the population), then the chance of each child's having PKU is one half:

From Liz	From Partner	Result
pah$^-$	*PAH*$^+$	Healthy (carrier)
pah$^-$	*pah*$^-$	Suffers from phenylketonuria

The inheritance of PKU is precisely like that of many other diseases caused by recessive mutations. Most metabolic abnormalities show recessive inheritance because a single copy of a gene encoding an enzyme that carries out a step in metabolism is generally sufficient to provide enough of the protein for normal life; the loss of half of the amount of enzyme due to one bad copy of the gene usually has no ill effect. Fortunately for us, most of these diseases are rare because it's not often the case that both parents will carry a mutation in the same gene. Some mutations are incredibly rare; for example, only about twenty cases of a type of congenital insensitivity to pain with anhidrosis (inability to sweat), caused by a mutation in a gene that codes for a protein that stimulates nerve growth, have been reported. Others, however, are not so rare. Gaucher disease—whose symptoms include joint degenerations, bone fractures, and bleeding problems—occurs in about one in fifty thousand people and is caused by two defective copies of the gene encoding an enzyme called beta-glucosidase.

The most common recessive disease among Caucasians is cystic fibrosis, which is caused by a mutation in a gene known as *CFTR* (which codes for a protein called the cystic fibrosis transmembrane conductance regulator). About 3 percent of Caucasians carry a mutation in the *CFTR* gene.

Recessive inheritance of disease can also occur for genes on the X chromosome, as we saw in chapter 5 for the *IL2RG* gene, whose mutation causes SCID, severe combined immunodeficiency syndrome. Because females have two X chromosomes, both copies of a gene on this chromosome must be defective for a recessive disease to be manifest, just as for genes on any of the other chromosomes. But males, with their lone X chromosome, have no backup copy of the genes it carries, so a mutation in a gene on the X can cause major problems for boys. For example, Duchenne's muscular dystrophy is due to a mutation in a gene on the X chromosome that specifies a protein important for muscle function. About 1 in 3,500 boys inherit the defective gene from their carrier mothers (their fathers give them a Y chromosome).

Young Carol, back in 1920s China, was afflicted with phenylketonuria. Her mother, Pearl S. Buck, of course knew nothing of this disease, whose name would not be coined for another two decades. Buck knew only of her child's failure to develop normally. She visited institutions all over the United States to find a home for Carol, eventually deciding on the Training School in Vineland, New Jersey, because the staff treated the children with dignity and carried out research on the causes of mental disabilities. After leaving Carol at the school, Pearl Buck did not see her again for three years.

To earn money for Carol's care, Buck obtained five hundred dollars from the Presbyterian Mission Board in New York to write a children's story about missionaries. This work led to other writing, and in 1931, two years after Carol began life at Vineland, Buck published *The Good Earth,* a tale of peasant life in China, for which she won the Pulitzer Prize. In 1938, Buck became the first American woman to win the Nobel Prize for Literature.

In *The Good Earth,* Wang Lung and his wife O-lan have a baby girl who is profoundly retarded. This would not be the only work of Buck's to feature children with mental deficiencies, but it was not until 1950, with the publication of *The Child Who Never Grew,* that she revealed her daughter's—and her own—history. Buck's efforts on behalf of children with mental retardation spurred others to action, including Eunice Kennedy Shriver, sister of President John F. Kennedy, who wrote about their sister Rosemary, and Dale Evans Rogers, the wife of the cowboy actor Roy Rogers, who wrote about their child with Down syndrome.

In 1960, a pediatrician involved in PKU research visited Pearl Buck at her home in Pennsylvania, and asked her to smell a vial containing phenylacetate crystals, which produce the odor of stale urine from PKU patients. Buck immediately remembered that Carol produced this same characteristic odor, indicating that her mental disability was indeed due to PKU.

Carol Buck lived at Vineland for more than sixty years; she died of cancer in 1992 at the age of seventy-two. Progress in medical research over her lifetime has largely answered the questions that so puzzled Pearl Buck and the Egelands. Why did these children get phenylketonuria? Because their personal DNA codes included two copies of a defective *PAH* gene, one inherited from each of their unsuspecting carrier parents. How did they come to be so profoundly disabled? Because the buildup of phenylalanine in the blood inhibits development of an infant's nervous system. What can be done about the disease? A diet designed to minimize the intake of phenylalanine heads off the cognitive deficiencies and allows those with the disease to lead nearly normal lives. Because these questions could not be answered in 1920, the world got one of its literary treasures.

8 When One Gene Is Too Much: At Risk for Huntington's Disease

Mike O'Brien, thirty-nine, and his brother Chris, thirty-two, wanted to become the first American brothers to reach the summit of Mount Everest together. Experienced climbers, they had previously scaled twenty-five mountains together, including Mount Kilimanjaro in Tanzania, Mount Ranier in Washington, and Cho Oyu, the sixth-highest mountain in the world, in Tibet. On the website they set up to document their Everest expedition, they described themselves as "the 4th and 7th children born into the O'Brien clan of seven. [Their parents] passed on to their children fine table manners, swimming talents, [a sense] of humor, good looks and a strong sense of pride."

On May 1, 2005, Mike called his girlfriend, Rebecca Stodola, at their Seattle home, from camp 2, above their base camp. "He sounded like a kid on Christmas—his voice sounded breathy with excitement," Stodala said. "He said, 'We're ready. We're psyched, and if the mountain lets us, we'll do it.'" But shortly after the call, Mike and Chris decided to descend to base camp because camp 3, farther up the mountain, wasn't yet ready.

Starting off from camp 2 at 21,200 feet, they reached camp 1, at 19,500 feet, and began to traverse the Khumbu Ice Fall on their way down to the base camp. Around 1:45 p.m., Blair Falahey, who was walking behind the O'Briens, shouted to Chris that his brother had fallen into a crevasse. Mike lay on his back about twenty feet below the trail. He said that he thought he had broken his leg and damaged some ribs, but he could move his head and all his limbs.

Blair and Chris climbed into the crevasse and tried to make Mike comfortable. They put a hat, sunglasses, and sun cream on him, moved a sleeping mat underneath him, and propped his head up. But Mike began asking, "How long for oxygen to arrive. It's hard to breathe." Two other climbers raced back toward base camp to get help and oxygen. Mike's condition was stable for about thirty minutes, but then rapidly deteriorated as his breathing difficulties increased. When he stopped breathing, Blair and Chris began CPR and rescue breathing, and when oxygen arrived, they gave it to him at the highest possible flow rate. But Mike never took another breath, and at around 4:30 p.m. they gave up their efforts.

The assistant team leader, Mark Merwin, arrived on the scene around 6:30 p.m. and examined the terrain where the accident had occurred. He surmised that Mike, like most other climbers at that point in the descent, had not been "clipped to the fixed line with his safety when he fell. If he had clipped the rope I think the rope would have arrested his fall, or we would have seen a broken rope at the scene of the accident. We did not find the rope broken. Nor was his safety line nor carabiner damaged in any way, as if it might have broken in a fall. . . . I think it is possible that he simply tripped and fell while unclipped to the rope." The weather had been unusually warm, which may have caused snow to become impacted in Mike's crampons and loosen his footing.

Mike "lived adventure to adventure," Stodola said. She also said that he had decided to "live life as if he might not see forty." In fact, Mike knew full well that an early death was a fifty-fifty possibility for him. The brothers' website contained the information that David and Alice O'Brien had passed on to their children, in addition to their other gifts, "a terrible genetic disease called Huntington's." The boys' mother had died of the disease at fifty-nine; their sister Diane had died at the age of thirty-nine, after she took a fall while still in the early stages of the disease; and their youngest sister, Alice, had recently begun to develop symptoms. Mike and Chris had taken on the Everest expedition as a fundraising project: they hoped to raise $100,000 for the Hereditary Disease Foundation, which supports the search for a cure for Huntington's disease.

Huntington's disease was first described in the medical literature in 1872 by George Huntington, a twenty-two-year-old American doctor just one year out of Columbia University Medical School in New York City. Huntington drew on observations of his father and grandfather, also physicians, who described to him the involuntary shaking they had observed in some of their patients with a disease called "St. Vitus Dance." Its most well-known sufferer may be the American folksinger Woody Guthrie, who succumbed to the disease in 1967 after spending many years in mental institutions and hospitals where he had been misdiagnosed as an alcoholic. Although he knew nothing of Mendel's work explaining how traits are inherited, Huntington accurately described the genetic mode of transmission of this horrible disease.

The symptoms of Huntington's disease usually become apparent between thirty and fifty years of age, and they foreshadow a long and difficult path that leads to death. Nerve cells in certain areas of the brain degenerate, resulting in uncontrolled movements as well as cognitive and behavioral problems. People in the early stages of the disease are often irritable, depressed, anxious, and aggressive. They may have difficulty with mental processing and decision-making. Sometimes the first signs of the disease are involuntary movements or mild clumsiness or balance problems. Eventually, sufferers require major assistance with the activities of daily life, and many become unable to communicate.

About one in ten thousand people get the disease. Unlike the recessive disorder phenylketonuria, which results only if both copies of a defective *PAH* gene are inherited, Huntington's is a dominant genetic disease: a single copy of the abnormal gene (which we'll call HD^*) is sufficient to cause the disease. Any individual whose personal DNA code includes one abnormal HD^* gene will eventually succumb to the disorder. People destined to die of the disease typically have as well a copy of the normal gene, which we'll call hd^+, so they are HD^*/hd^+. Those who will not get the disease are hd^+/hd^+.

Males and females get Huntington's disease in equal numbers because the Huntington's disease gene is not located on either the X or Y sex chromosomes. Since Huntington's disease is a dominant disorder, each child of a parent with the disease has a fifty-fifty chance of inheriting the abnormal gene and developing the disease. Each also has a fifty-fifty chance of inheriting the normal gene. If a mother has Huntington's disease (HD^*/hd^+), on average, half of her eggs will carry the abnormal HD^* gene for Huntington's disease. She inherited that gene (and, therefore, the disease) from whichever of her parents had it. But half of her eggs will carry the normal version of the gene inherited from her parent who was free of Huntington's. Each time an egg is fertilized to make an embryo, there is a 50 percent chance that the egg will contain the abnormal version of the *HD* gene and a 50 percent chance that it will carry the normal version.

(Because the *HD** mutation is rare in the human population, virtually everyone who has the mutation mates with someone with the normal version of the gene.)

From Mom	From Dad	Result
*HD**	*hd*+	Suffers from Huntington's
hd+	*hd*+	Healthy

In the same way, each child whose father will succumb to Huntington's disease (*HD**/*hd*+) also has a fifty-fifty risk of getting the disease. Half of the sperm of the father will carry a normal version of the *HD* gene and half will carry an abnormal version of the *HD* gene.

From Mom	From Dad	Result
hd+	*HD**	Suffers from Huntington's
hd+	*hd*+	Healthy

Imagine you are flipping a quarter 100 times. Each time you flip, you have a fifty-fifty chance of getting either heads or tails. It's the same thing with the genetic flip of the coin. Each child born to either a mother or father with Huntington's disease has a 50 percent risk of having the disease. It does not matter which embryos have come before or will follow; with each pregnancy, the statistical chances are the same. In some families, just by bad luck, all of the children are destined to develop the illness; in other more lucky families, none do. The fate of one child has absolutely no bearing whatsoever on the fate of any other.

If you're an at-risk individual because your dad or mom developed Huntington's, what determines whether you inherited the normal chromosome, or were fated to contract the disease by inheriting the abnormal version? Although scientists can now analyze individual sperm cells and determine whether the Huntington's disease gene in them is the normal or abnormal version, to an egg cell the billions of sperm are indistinguishable. The decisive event of fertilization is random, one of the many accidents of the universe we inhabit. No force guides one sperm cell or another to a successful fertilization, any more than the heads or tails of a single coin toss is preordained. We are the products of numerous chance events that determined our personal DNA code. Neither the child born with a severe genetic affliction nor the parent is to blame.

The Huntington's disease gene lies near the tip of chromosome 4. Identification of the approximate location of the gene in 1983 electrified human geneticists because it was the first success of a new approach to map disease genes based on the knowledge of the human genome that was starting to accumulate. The discovery relied heavily on the work of Nancy Wexler, a neuropsychologist at Columbia University, and her colleagues.

Huntington's disease is prevalent in the communities around Lake Maracaibo, in Venezuela; up to half the residents of some small villages are ravaged by it. It's not unusual to see villagers there walking around with legs and arms flailing about in wild, uncontrolled movements. This extreme prevalence of Huntington's is due to a woman with the disease who lived there about two hundred years ago and gave birth to ten children. Because of the isolation of these communities, the descendants of this woman now make up much of the current population, a phenomenon known as a founder effect. The advantage to geneticists of such a population is that all those with the disease carry the same mutation, and the contribution of the environment is minimal because all the residents share the same living conditions.

Wexler and her colleagues constructed an extensive family tree of the thousands of Huntington's sufferers in this region. She took the first of many trips to the Lake Maracaibo region in 1979, taking blood samples and skin biopsies from the residents and tracking the progression of the disease in those showing symptoms. Her older sister, Alice, who accompanied Nancy in 1983 and 1984, wrote in her book about the disease, *Mapping Fate*, "Nancy is physician, nurse, ethnologist, psychologist, diplomat, photographer, neurologist, geneticist, and general all rolled into one." Collection of these samples in the primitive conditions in the Venezuelan villages required overcoming numerous logistical hurdles, like timing the days for blood draws to coincide with the departure dates of members of the team so they could hand-carry the tissues back to the United States. These samples proved crucial both for localizing the gene to a region of the human genome and then, ten years later, for identifying the DNA sequence of the gene.

The *HD* gene encodes a protein called Huntingtin. The change in the protein caused by the mutation in the HD^* gene is unusual. In the normal (hd^+) gene, there is a stretch of DNA where the bases CAG, which specify the amino acid glutamine, are repeated: CAGCAGCAGCAGCAGCAG and so forth. Individuals with the normal (*hd*$^+$) version of the gene have between seven and thirty-five copies of CAG at this point in the gene, resulting in seven to thirty-five glutamines all in a row in the Huntingtin protein. The *HD** (abnormal) versions of the gene have forty or more copies of CAG, resulting in a longer run of glutamines in the Huntingtin protein. It's not clear why, but those extra glutamines in Huntingtin cause the protein to poison brain-cell function, ultimately leading to an early death. (Some people with thirty-five to thirty-nine copies of CAG in the gene may develop the disease; others will escape it.)

Why do we term the mutation that causes Huntington's disease "dominant"? For recessive mutations like the one that causes phenylketonuria, the amount of protein produced by a single functional copy of the *PAH*$^+$ gene is sufficient for its normal function in development. Both copies of the gene must be defective for the disease to occur:

From Mom	From Dad	Result
Functional *PAH*$^+$	Defective *pah*$^-$	Healthy (carrier)

Huntington's disease, however, occurs even though one of the copies of the gene produces normal Huntingtin protein in each cell of the body:

From Mom	From Dad	Result
Altered *HD**	Normal *hd*$^+$	Huntington's disease

The altered Huntingtin protein with those extra glutamines is toxic, possibly because those amino acid residues cause the protein to associate abnormally with other proteins in the cell, resulting in clumps of proteins that effectively gum up the works of the cell (a more precise description of the phenomenon can't be given because its basis is not well understood). Certain brain cells are especially susceptible to this protein clumping, causing the disease to be manifested largely as a neurological disorder.

Dominant mutations like these are called "gain-of-function" mutations because they confer on a protein an activity it doesn't normally have (in this case aggregation of Huntingtin with other proteins):

From Mom	From Dad	Result
Altered (gain-of-function) *HD**	Normal *hd*$^+$	Huntington's disease

Recessive mutations like those in the *PAH* gene typically cause a "loss of function" of the protein they affect.

From Mom	From Dad	Result
Functional PAH^{+}	Defective (loss-of- function) pah^{-}	Healthy (carrier)

Since about 1 in 10,000 people carry the HD^{*} mutation, the likelihood that an individual will inherit two copies of a defective HD^{*} gene is about 1 in 100 million (1/10,000, the chance that Mom carries the mutation, times 1/10,000, the chance that Dad carries the mutation). Such individuals are found around Lake Maracaibo because of the large number of people carrying the abnormal form of the gene.

Ever since the *HD* gene was discovered in 1993, people like Mike O'Brien who are at risk of the disease can take a genetic test to find out if the abnormal gene is part of their personal DNA code and learn whether or not they are destined to develop symptoms of Huntington's disease. Genetic counselors isolate DNA from cells in a tissue sample (obtained by swabbing the inside of the cheek or drawing a bit of blood), and determine the precise number of CAG repeats in each of the two copies of the gene. There is no indication that Mike ever took such a test, or that his DNA was tested after he died. His sister Meghan concealed her true identity because of concerns about insurability in the event the test showed that she had inherited the abnormal gene. But she did not receive the disease-causing mutation, and she is now raising a family, having escaped from under the dark cloud of Huntington's disease.

The specter of taking a genetic test to establish the likelihood of falling victim to a progressive, incurable disease like Huntington's can be overwhelming to those at risk, whose worst fears may be confirmed by the test. Knowledge that the abnormal gene has been inherited brings a considerable penalty, removing that fifty-fifty probability and replacing it with the certainty of a slow, agonizing death. On the other hand, taking the test can be a source of relief if only the normal version of the gene is found, lifting the anxiety of a life in limbo and the daily stress of not knowing how much of a future to plan for.

For some individuals, the prospect of gaining peace of mind warrants the risk of certain knowledge of an awful fate. But even a test that provides good news can bring a heavy burden, including survivor's guilt over the other family members who are less fortunate, and the loss of a driving force to accomplish something that may have been motivated by one's genetic inheritance. Beyond the emotional stakes for at-risk individuals of learning their fate, there are potential practical risks, such as loss of employment or insurance, and even social stigma.

Nancy Wexler, the neuropsychologist who played a major role in finding the *HD* gene, and her sister Alice know the emotional stakes well. Their mother, Leonore, was stopped by a Los Angeles police officer in 1968 because she was walking erratically and was thought to be inebriated. But alcohol was not the problem; her poor balance was a consequence of Huntington's disease, which claimed her life a decade later. Their father, Milton, got together with the two girls, then in their early twenties, to tell them that their mother had "'a progressive, degenerative, neurological illness, that it often caused madness, that it was always fatal, and that both [of you] . . . have a fifty-fifty chance of inheriting the illness yourselves.' He went on, 'You know what you both said? "Fifty-fifty? That's not so bad." That took a terrific load off my mind.'"

The awareness that Leonore Wexler had Huntington's disease unearthed hidden family secrets—a not uncommon occurrence in such families. The sisters learned that their maternal grandfather and all three of their mother's brothers had succumbed to the disease. Her mother had watched the inexorable decline of her father. When she was just fifteen she had read a textbook of neurology and noted that it stated that only males inherit the disease. Hoping to find a cure for her older brothers before it was too late, she studied genetics in college and worked in the laboratory of the Nobel Prize–winning geneticist Thomas Hunt Morgan.

When a genetic test for the disease became available, Nancy and Alice Wexler had to confront their own ambivalence about the issue. Heated family discussions took place in 1984. Alice described her decision to forego the test: "'Talking about it concretely—moving from the realm of abstract possibility to planning the logistics of it—terrifies me. . . . The thought of learning that I carry that gene—that my brain is already deteriorating—is just too horrendous. I'm not sure that I could go on.'"

Their father, Milton Wexler, a psychoanalyst who treated many notable Hollywood figures, founded the Hereditary Disease Foundation in 1968 in an effort to uncover the cause of the disease that would kill his wife and that threatened his daughters. He vowed to use the funds of the Foundation, for which his daughter Nancy now acts as president, to fund treatments and cures for this horrific illness, and he dedicated the next forty years of his life to finding the gene and making progress toward a cure. This is the organization for which Mike and Chris O'Brien were raising funds with their Everest expedition. A year after Mike O'Brien's untimely fall on the mountain, his family and friends succeeded in topping their $100,000 goal. But we are still awaiting a cure for Huntington's disease.

9 Genes to Remember: The Growing Burden of Alzheimer's Disease

Margarita Carmen Dolores Cansino began her life in Brooklyn, New York, in 1918, at the end of the First World War. The daughter of a showgirl from Washington, D.C., and a flamenco dancer from Madrid, Cansino was destined to go on to a celebrated career punctuated by tragedy. Intensely shy and quiet as a girl, she was kept out of school by a domineering father (who also sexually abused her, according to one biographer) so that she could perform Spanish dances in Mexican casinos and on gambling boats. At the age of eighteen, Cansino married another controlling man, a thrice-divorced lounge lizard in his forties who viewed her more as an investment than a wife. To make the investment pay off, her husband willingly pushed her into the arms of any man in Hollywood who might help get her into films. To improve her chances of stardom, Cansino changed her hairline, her hair color and her name, transforming herself into Rita Hayworth.

Hayworth became instantly recognizable as one of the most glamorous screen stars of her era. She made a series of musicals with Fred Astaire and Gene Kelly, as well as erotically charged films such as *Blood and Sand, Gilda,* and *The Lady from Shanghai.* A photograph of her kneeling on a bed in a negligee of satin and lace that appeared in Life magazine in 1941—sent to millions of American soldiers—was rivaled in popularity only by the one of Betty Grable in a white bathing suit. When Orson Welles saw the photograph while filming in South America, he told his friends that upon his return to the United States he would marry Hayworth, even though the director of *Citizen Kane* had not yet set eyes on the actress in person. And marry her he did.

Likely a result of her painful childhood, Hayworth suffered through difficult relationships with her five husbands and her many lovers, one of whom was the aviator and tycoon Howard Hughes. Outwardly the femme fatale, inwardly Hayworth remained a vulnerable little girl. Referring to her role as the eponymous sexual temptress Gilda in the sizzling film noir costarring Glenn Ford, she remarked, "Men go to bed with Gilda, but wake up with me."

Hayworth's career spiraled downward after she turned forty. It wasn't that her looks betrayed her or that she could no longer hold an audience. No, she was losing her mind. The cause was likely her possession of a different base at one particular position in her personal DNA code—say, where she might have had a T rather than the C that most people have there. That single difference in Hayworth's DNA would result in the production of a single altered protein, and it likely was this protein that ended her career.

On November 25, 1901, Auguste Deter, the fifty-one-year old wife of a railway worker, arrived at the Municipal Asylum for the Insane and Epileptic in Frankfurt-am-Main, Germany. Showing unusual symptoms of delusions and hallucinations, she was referred the following day to a research-oriented physician, who recorded their conversation.

"What is your name?" he asked her.
"Auguste."
"And your surname?"
"Auguste."
"What is your husband's name?"
"I think . . . Auguste."
"Your husband?"
"Oh, yes."
"How old are you?"
"Fifty-one."
"Where do you live?"
"Oh, you were so nice at our house."
"Are you married?"
"Ah, I am so confused."

Auguste's husband related the mystifying course of her illness. About eight months earlier Auguste had begun to believe that her husband was being unfaithful. She also began forgetting things, messing up her cooking, and wandering aimlessly through the house. These behaviors were soon followed by paranoia, fears of dying, and increasing forgetfulness.

At the Frankfurt clinic Auguste's condition continued to deteriorate. She became hostile and violent, striking other patients in the face. At the same time, Auguste grew less and less animated, so that three years after her admission she simply lay curled up in bed, completely dazed. By the following year, she was totally silent. In April of 1906, five years after the appearance of her first symptoms, Auguste Deter, physically and mentally only a shadow of her former self, died in the clinic of septicemia caused by bed sores.

Decades later, however, she would become one of the most famous patients in the annals of medicine. Auguste's autopsy samples were sent to a well-known lab, the Cerebral Anatomical Laboratory in Munich, where the physician who had first interviewed her had since moved. He analyzed Auguste's brain with the latest procedures and the most powerful microscopes of the day. Two findings stood out in his analysis. First, inside the brain cells—called neurons—he observed bundles of thick fibers that came to be known as neurofibrillary tangles. Second, outside the cells he saw deposits of an abnormal substance that came to be known as amyloid. At a conference in late 1906, Dr. Alois Alzheimer presented his paper on "profound dementia in a young adult." He concluded, "I have just presented a clearly defined and hitherto unrecognized disorder." By all accounts, Alzheimer's presentation had no impact on the audience of psychiatrists who heard it.

Because Auguste displayed symptoms at only fifty-one years of age, her disease—dubbed Alzheimer's disease a few years later by Alzheimer's mentor, Emil Kraepelin, the father of modern psychiatry—was classified as a presenile dementia, defined then as occurring before sixty years of age. This disorder stood in contrast to the much more common senile dementia displayed by older patients. This distinction between dementia in a younger person and dementia in the aged would confound the field for half a century, during which time Alzheimer's disease attracted hardly any attention from researchers.

Whereas presenile dementia showed characteristics that caused it to be considered a disease, senile dementia was thought to be to the result of the normal deterioration of old age. Instead of giving credence to the idea of a physical illness, psychiatrists fostered a model of dementia in the aged that pinned the cause on some combination of personality, emotional trauma, mandatory retirement, social isolation, and the breakup of the family. In the 1940s and 1950s they linked social pathology to brain pathology, viewing the former as the cause of the latter. To psychiatrists of that era, it was all nurture, and no nature.

By the 1970s, however, society had come to appreciate that people should not be discriminated against on the basis of their advanced age any more than because of their race or sex. As a consequence, the stereotype of aging individuals as inevitably succumbing to dementia began to fade. Meanwhile, on the scientific front, Martin Roth, Bernard Tomlinson, and Gary Blessed at Newcastle University, in the U.K., demonstrated that the degree of dementia in a person correlated best not with age but with the number of deposits of amyloid and neurofibrillary tangles observed in the brain, suggesting that specific pathological changes, and not the general aging process, were the culprit. Because the clinical and pathological manifestations of the early-onset (presenile) and late-onset (senile) dementias were the same, both disorders were classified as Alzheimer's disease.

These findings led to a much greater realization of the burden of Alzheimer's disease, which until then had been thought to be relatively small. In 1976, Robert Katzman of the University of California in San Diego made the bold estimate that Alzheimer's disease afflicted 1.2 million Americans and accounted for sixty thousand to ninety thousand deaths each year. Those numbers startled the public and galvanized the research community, stimulating big increases in funding for research into the causes of the disease and the development of therapies. The numbers from thirty years ago pale compared to those of today: about 5 million Americans currently suffer from this horrible disease; by 2050 it is expected to be between 11 million and 16 million.

Inside the cell, neurofibrillary tangles of protein fibers; outside the cell, deposits of amyloid protein. The actual makeup of those defining structures and the process by which they form remained mysterious for eighty years after Alois Alzheimer's examination of Auguste Deter's brain. Indeed, whether the tangles or deposits cause the disease or merely appear as a byproduct of the real toxic event remained an open question for much of the last century. Maybe dying cells in the brain accumulate broken-down bits of worn out proteins, and these are what form these two structures, long after the cells have quit functioning for completely different reasons.

But over the last two decades, Alzheimer's disease research has made remarkable leaps, and the geneticists can take much of the credit. Their studies of rare hereditary forms of Alzheimer's disease, although comprising less than 1 percent of all cases, have provided insight into the more prevalent so-called "sporadic" cases (whose cause is unknown). These studies have revealed the pathways that produce the tangles and deposits seen in the common forms of the disease. The proteins implicated by these rare familial cases of Alzheimer's disease provide targets for developing medication to treat Alzheimer's.

Before plunging into the genetics of Alzheimer's disease, we first need to consider its cast of proteins. In 1984, George Glenner, a pathologist at the University of California in San Diego, identified the small protein present in the amyloid deposits, which he named amyloid-beta. Not long after that, amyloid-beta was found to be derived from a much larger protein called amyloid precursor protein that sits on the cell surface with just a short tail of amino acids protruding into the cell. The protein gets clipped in two places to produce amyloid-beta. The scissors that do the clipping are two other proteins that researchers named presenilin-1 and presenilin-2.

Meanwhile, geneticists collected blood from families around the world whose members exhibited the early-onset form of Alzheimer's disease. They subjected the DNA of these family members to intense scrutiny, using new powerful tools for DNA analysis that were being developed.

In 1987, the gene for amyloid precursor protein was identified. Intriguingly, it lies on chromosome 21, an extra copy of which causes Down syndrome. That is significant because people with Down syndrome typically exhibit symptoms of Alzheimer's by age forty, possibly because of extra amyloid-beta protein produced from their extra amyloid precursor protein gene on their extra chromosome 21.

The geneticists eagerly examined their Alzheimer's family pedigrees, hoping to find mutations that cause the disease in the newly discovered amyloid precursor protein gene. In 1991 they found one in a British family with a rare form of Alzheimer's. A few other mutations in this gene have since been identified, most of which change amino acids near the sites in the protein that get clipped by the presenilin scissors in the process of generating amyloid-beta. The key fact about these mutant versions of the gene is that they lead to an abnormal accumulation of amyloid-beta, and therefore implicate amyloid-beta as a cause, not just a consequence, of the disease.

But mutations in the gene encoding amyloid precursor protein are the least frequent cause of familial Alzheimer's disease. Most familial cases are due to mutations in the genes encoding the presenilin-1 and presenilin-2 scissors. The first of these was found in the mid-1990s after an enormous amount of work on pedigrees of families with Alzheimer's disease (the use of pedigrees will be discussed in more detail in chapter 13). Some mutations in the presenilin genes cause the scissors to be hyperactive, resulting in the production of excess amyloid-beta, strengthening the case that amyloid deposition is a cause of Alzheimer's disease.

Geneticists conclusively demonstrated that amyloid-beta contributes to Alzheimer's disease, but that still left the question of the cause of the neurofibrillary tangles that accumulate in neurons in the brain of people with the disease. The major component of the tangles is a protein known as tau, part of a network of protein cables that move materials around the cell, much as the cables of a chairlift move skiers up the mountain. Since neurons are large cells with long extensions, they are especially reliant on these cables to move cellular materials long distances.

In 1998 geneticists identified disease-causing mutations in the gene that encodes tau protein, but not in patients with Alzheimer's disease. Rather, they found the tau mutations in patients suffering from a neurological disorder called frontotemporal dementia and Parkinsonism linked to chromosome 17, or FTDP-17 for short. People with this disorder have intracellular deposits of tau, but no deposits of amyloid-beta. The hypothesis that tau plays a role in this disease supports the idea that altered forms of tau cause neurodegeneration. But they do not seem to be a cause of Alzheimer's disease.

The mutations in the genes that encode amyloid precursor protein, the presenilin scissors, and tau shed much needed light on the etiology of Alzheimer's disease. But the laurels will go to those who identify mutations that contribute to the common, sporadic form of the disease. Many such mutations must exist, but so far only one major genetic risk factor is known: the gene encoding apolipoprotein E (apoE), a protein that shuttles cholesterol around the body, including to the brain. The particular form of apoE specified by your personal DNA code influences your disposition to Alzheimer's disease, but it doesn't strictly determine your fate: people carrying one copy of the apoE4 form of the gene are only about three times more likely to get the disease. Those with two copies are nine times more likely, but some people with two copies live into their eighties and never get the disease. Furthermore, Alzheimer's disease is common, and apoE is only one risk factor; only about half of Alzheimer's sufferers have the apoE4 form of that gene.

Rita Hayworth lost her career and her life to Alzheimer's disease. Like many people, both of us have seen loved ones suffer this awful disease, which takes a heavy toll on the family as well as the patient. Most Alzheimer's researchers think that amyloid-beta is one of the culprits that rob victims first of memory and, eventually, of almost all other thought processes. The accumulating amyloid-beta somehow gums up the works of cells in the brain, eventually killing them.

The early stages of Rita Hayworth's disease brought mood swings, tantrums, and violent outbursts that were attributed to alcohol abuse. The memory deficits soon became apparent. When Hayworth was tapped to replace Lauren Bacall in the Broadway production of *Applause,* she discovered she couldn't remember a whole play's worth of dialogue, and she had to pull out of the production. By 1971, the fifty-three-year old actress had trouble retaining even short bits of dialogue. During the filming of *Wrath of God,* she had to be fed one line at a time. Some years later Orson Welles ran into her at a hotel and went over to her and kissed her. "My blood ran cold," Welles recalled, as he realized that Hayworth had no idea who he was.

The early onset of Rita Hayworth's Alzheimer's disease indicates that it was likely caused by a mutation in her personal DNA code: a substitution of one base-pair for another that changed one of the base triplets of a gene, causing a different amino acid to be inserted at that position in a particular protein chain. That amino acid change at that particular spot in the protein altered its chemical properties and disrupted its structure, causing it to kill neurons. We don't know which of Rita Hayworth's genes suffered that fateful mutation—perhaps it was one of the presenilin genes—but it's likely that a change in only one of her 6 billion base-pairs led to the buildup of the amyloid-beta that shortened her dazzling career and prematurely ended her life.

When Hayworth died in 1987 at the age of sixty-eight, a fellow actor was occupying the White House. "Rita Hayworth was one of our country's most beloved stars," Ronald Reagan said. "Glamorous and talented, she gave us many wonderful moments on the stage and screen and delighted audiences from the time she was a young girl. . . . Nancy and I are saddened by Rita's death. She was a friend whom we will miss."

The president also praised Hayworth and her daughter, Princess Yasmin Aga Khan, for publicly confronting Alzheimer's disease. "The courage and sincerity shown by Rita and her family have done us a great public service by calling the world's attention to a disease that all of us hope will soon be curable." Alas, the cure did not come soon enough for Mr. Reagan: he himself succumbed to the disease in 2004, ten years after telling the American public of his own diagnosis.

Unlike Hayworth's disease, the Alzheimer's disease that we typically confront in our parents or grandparents, our aunts or uncles, does not strike the young. The Alzheimer's Association estimates that about one fifth of Americans who reach the age of seventy-five will suffer from the disease in their next decade of life; for those eighty-five and older it's more than two fifths.

Research in recent years has provided clues to the risk determinants, protective factors, and possible targets for drugs against this disease. The greatest risk factor is obvious: age. Before life expectancy in modern societies reached nearly eight decades, Alzheimer's disease had little impact. As more of us enjoy longer lives, more of us suffer from this devastating disease.

Another risk factor is also clear: one's parents. People with a parent or sibling with the disease are two to three times more likely to develop it themselves.

Several environmental risk factors have also been identified. Severe head injury increases the likelihood of getting the disease. Boxers who suffer years of blows to the head often end up with dementia. Other factors include the same ones that contribute to vascular disease—a high-cholesterol diet, high blood pressure, and diabetes.

Protective factors are also being identified. An intriguing one is high educational attainment, suggesting that a cognitive reserve, created over a lifetime of challenging mental activity, protects against Alzheimer's disease. People who are on statin drugs (for treatment of high cholesterol) or who take nonsteroidal anti-inflammatory drugs (for treatment of pain) seem to have a decreased risk in some studies, though these therapies have yet to prove useful in drug trials for those who already have the disease.

Since Alzheimer's disease generally develops in a person's seventh or eighth decade of life, a drug that staves off the symptoms for even a decade would be a blockbuster. Advances in understanding the basis of Alzheimer's disease have provided several proteins that seem like good targets for a drug. These include the presenilin proteins that cleave the amyloid precursor protein, proteins that contribute to the formation of amyloid-beta deposits, and proteins that remove amyloid-beta from the brain. Other approaches focus on tau, apoE, or growth factors active in the brain.

Rita Hayworth did not lose her memory, judgment, and language to alcohol abuse, as was first suspected. No, she succumbed to a neurodegenerative disease first diagnosed by Dr. Alois Alzheimer in Munich. Millions of other aging individuals who demonstrate failing memory are showing the effects of this disease, not of the aging process. Millions more will be affected before a cure is found.

10 Blaming Our Genes: The Heritability of Behavior

It's easy to accept that human disorders such as phenylketonuria or cystic fibrosis or Huntington's disease have a wholly genetic basis. And you likely have no problem believing that your risk of being afflicted with an illness such as heart disease, diabetes, or colon cancer is influenced by your personal DNA code. The question of heredity becomes more complicated, however, when we consider complex behaviors. Is your chance of having a sunny disposition affected by your genes? How about if you are a pessimist—is pessimism an inherited trait? What if you're an early-morning or a late-night person, or compulsively neat, or emotionally unable to connect with others—could genetic differences contribute to those traits?

How much of our temperament is due to our genes? How much of our intelligence—our ability to learn and remember, to acquire language, to read and spell—is a function of our DNA code? Will a tendency to become addicted to drugs, nicotine, or alcohol be passed along to our children? And what about psychiatric disorders such as schizophrenia, bipolar disorder, and depression? Are those due to our circumstances, or to our genetic constitution?

The answer is: both. Nature and nurture, our genetic endowment and our life situation and experiences, determine our behavior. But we suspect that the contribution from genetics may be more than most of us imagine.

Consider, for example, Susan Middlebrook, of Colchester, Vermont. She usually gets seven or eight hours of sleep each night, but rather than going to bed at 11:30 p.m., after the late-night news, and getting up at around 7 a.m., Middlebrook goes to bed early, very early . . . like at 6:30 p.m. And between 1:30 and 3 a.m., just when you're deep in your dreams, Middlebrook is rarin' to get up and go. "The net result is you can feel very isolated," she told John Roach of National Geographic News. "Who wants to party at three in the morning? Nobody I know, and I'm not headed to the local bar to see who's there." Instead, Middlebrook gets started on her morning chores long before your dreams are over.

Middlebrook suffers from familial advanced sleep phase syndrome (FASPS), the consequence of which is that her sleep patterns are way out of sync with the norm. To establish the norm, and the abnormal exceptions, sleep researchers ask people a series of questions about their sleep habits: At what time of day do you feel your best? How easy is it for you to get up in the morning? At what time would you go to bed if you were completely free to plan your day? At what time in the evening do you feel drowsy and begin to doze? Do you consider yourself a "morning" or an "evening" person? (The questionnaire was developed by a British scientist, James Horne, and a Swedish scientist, Olov Östberg.) Answers to these and other questions produce a score that lies on a scale running from extreme "eveningness" to extreme "morningness"; most of us fall somewhere in the middle of a broad bell-shaped curve of scores, but FASPS patients lie at the far "morningness" end of the scale, scoring higher than more than 99 percent of other respondents.

What makes Middlebrook's syndrome so interesting is that it's the result of an abnormality in her biological clock, known as the circadian rhythm; ("circadian" means, literally, the cycle of one complete day, from the Latin *circa-*, "cycle," and *dies*, "day"). This circadian clock controls not just sleeping and waking but also metabolic, physiological and behavioral processes such as heart rate, hormone levels, blood pressure, mood, and alertness. Our biological clocks are reset each day by exposure to sunlight, thus keeping us in synchrony with our surroundings. But Middlebrook's internal clock runs on a faster cycle, advanced by about four hours relative to most everyone else's, so that she is ready for bed when her neighbors are ready to sit down to dinner.

Middlebrook's sleep-wake cycle may seem unusual to you—it is far outside the statistical norm—but around the Middlebrook home it's about as unusual as oatmeal for breakfast: two of her three sisters, one of her parents, and her own child all keep the same odd hours. This syndrome is clearly due to a variation present in the personal DNA code of these family members (the word "familial" in the name of the syndrome might have tipped you off to this), and not to any environmental cause. Although FASPS affects only about three people in one thousand, its genetic basis was well worth deciphering. Findings about this syndrome may help to explain other more common sleeping disorders such as insomnia and narcolepsy, or lead to treatments for jet lag or seasonal affective disorder (SAD, in which the moods of sufferers are strongly affected by lack of sunlight in the winter), and they may provide health support and guidance for the 20 percent of workers who work the nightshift. Understanding human circadian rhythms may even enable drug treatments to be timed for better effect, or suggest ways to reduce nighttime auto accidents.

Now consider Jean, a twenty-four-year old nurse who became obsessed with cleaning and washing. As related by Peter McGuffin and David Mawson, two psychiatrists in London to whom Jean was eventually referred, each day she would wash her hands sixty to eighty times and spend twelve hours disinfecting her house; she went through twenty liters of disinfectant a week. She made her husband stop his sports activity because it brought dirt into the house, and she gave up sexual activity because it could cause contamination. When Jean was admitted to the hospital, her hands were "roughened, red, cracked and bleeding" from the constant washing.

Jean's identical twin, Jill, a social worker, lived apart from her sister and rarely saw her. Yet at the age of twenty-two, Jill began displaying behavior similar to Jean's. As soon as a meal was over she felt compelled to immediately wash the dishes and silverware, or else she would become severely anxious. Moreover, "the washing of the dishes and utensils had to proceed in a specific order, and failure to comply with the routine, or an attempt by others to relieve her of the task, provoked great discomfort." Jill established other washing rituals, and despite her best efforts to stop carrying them out, she spent increasing amounts of time on them, to the detriment of her social life and her work.

Both twins had normal childhoods, had been outgoing children, and were academically accomplished. None of their family members had received treatment for a psychiatric disorder, although their father was "fastidiously neat and orderly in his habits." The twins attended different universities; Jean obtained a degree in physics, Jill in history. Neither of them displayed any signs of mental illness before their obsessive-compulsive behaviors began. In fact, each of them learned of the other's similar problem only after Jean began treatment with the drug clomipramine, which together with behavior therapy alleviated her symptoms enough to allow her to return to nursing. Jill was not treated, but experienced a spontaneous remission of her symptoms after more than a year.

What caused these identical twins to develop a similar disorder at a similar time in their lives? Was it something about the way their parents brought them up that led to their symptoms years later? Was it some food they ate, or some toxin they were exposed to as children? Or was it because their personal DNA code is identical, and some combination of variations in the genes they both inherited caused them to develop an obsessive-compulsive disorder in their early twenties? Or did the disorder develop because of a complex interplay of these factors?

Consider, finally, Lewis, the son of a well-to-do couple in Pittsburgh. Lewis looked perfect at birth and appeared to be developing normally. As detailed by his mother, at eighteen months Lewis was almost saying words, but they weren't the typical words that most toddlers begin with, such as "Mommy" and "Daddy." Instead he babbled nonsensically. And he didn't respond to the usual games that toddlers enjoy, such as peek-a-boo or ring-around-a-rosie. At two years of age, he had no interest in some toys but was obsessed with others, and when he played with these he was oblivious to everything going on around him. He loved to climb and to swing and to bounce on a trampoline, which he did without fear. Yet sometimes he was afraid to step off a rug onto a hardwood floor, reacting as if the next step would be off a cliff. At night, Lewis took off all his clothes and slept on the floor wrapped in a blanket and surrounded by toy soldiers and matchbox cars. Lewis also had tantrums, during which he lay on the floor screaming, unable to indicate his problem.

At three he had yet to acquire any language and had little interest in communicating with his family. He babbled only to his toy soldiers. His doctors were no longer able to dismiss his problems as simply those of a "difficult" child or one in the throes of the "terrible twos." Lewis underwent a series of tests and the doctors concluded he had an autism spectrum disorder.

The diagnosis led to a variety of publicly and privately funded treatments, including speech and occupational therapy. Lewis made remarkable progress: he developed some language ability, and his behavior calmed down. He could hold his mother's hand, play peek-a-boo with her and kiss her good night, and occasionally even make eye contact. When he read his favorite book, Eric Carle's *The Very Hungry Caterpillar*, he said every word at the right time and mimicked the slurping sound of the caterpillar eating. But he still could not carry on a conversation. Lewis entered a special ed preschool class, with the hope and expectation that in a few years he could be integrated into a typical school class.

Autism is a complex developmental disorder that appears within the first three years of life. Its symptoms, which can range from mild to severe, are characterized by a lack of emotional contact with others, difficulty with verbal and nonverbal communication, and a restricted range of activities and interests. Children are born with autism or with the potential to develop it; they do not acquire the condition because of bad parenting. A fascinating feature of the chromosomes of some autistic children is that large chunks of DNA—millions or tens of millions of base-pairs—are found duplicated in some children and lost altogether in others. These regions can contain dozens of genes. The analysis of the personal DNA codes of these children may help pinpoint the specific genetic changes responsible for this complex and debilitating disease.

When a human population is measured for any characteristic, say height or weight, or blood pressure or cholesterol level, the measurements reveal a continuous spectrum between two extremes. When it comes to height, most men fall somewhere between former Charlotte Hornets basketball player Muggsy Bogues, five foot three inches, and seven foot six Yao Ming of the Houston Rockets.

It's no different for a complex behavior, such as how we deal with stress. Even very young babies display a broad range of responses, from those who cry easily to those who seem unbothered by almost any challenge. Similar patterns are observed in the extent or degree to which children are active or sedentary, whether they persist at pursuing a task or lose their focus, or the extent of their introversion or extroversion. Among adults, the amounts of alcohol, nicotine, or drugs that are consumed show the same broad distributions. Some part of these traits is likely due to genetics, and some is surely due to environmental components. But how much is due to each?

The measure of the amount of the genetic effect is termed "heritability." A trait or behavior that is wholly due to genetics has 100 percent heritability; one that is wholly due to the environment has zero heritability.

The environmental component could be due to factors shared among family members, such as the home environment, the food preferences of the family, or the level of pollution in the town where the family resides. Or it could be due to factors that are not shared, such as the specific set of friends or teachers of each family member, or unusual life events such as accidents, or diseases unique to each individual.

How do we measure, or quantify, heritability? It's actually pretty easy. All that is needed is a set of individuals whose genetic relatedness is known, and a measurement of some trait or condition for each individual.

The easiest of these studies to understand exploits the genetic similarity of twins. The personal DNA codes of identical twins are identical, because they develop from a single fertilized egg. That is, the DNA sequence of each gene in one twin is identical to its sequence in the other twin. Fraternal twins also shared the same womb at the same time, but have the same version of only about half of their genes (like any siblings), so the genetic contribution to a trait is twice as great in identical twins as it is in fraternal twins. Thus, the more similar a trait is in identical twins as compared to fraternal ones, the greater the genetic contribution to it is likely to be.

Another type of study compares siblings from the same parents (who have the same version of about half of their genes), to adopted siblings who are not genetically related but share the same surroundings. A genetic influence on a trait is apparent when biological siblings are more similar than adopted ones; an environmental influence is obvious when adopted and thus genetically unrelated siblings resemble each other more than they do other unrelated people who grew up in other families.

A third type of analysis uses family genetic studies that compare how often a disease occurs in a family in which one member is known to be affected to how often it occurs in the general population. Diseases with a genetic basis occur more frequently in members of a family that has one case—in other words, genetic diseases run in families.

With these ways to quantify heritability, let's examine some human traits. We'll start with intelligence—a particularly thorny issue. We'd like to believe that a roomful of books, Mozart playing in the background, and a nurturing set of parents will put any infant on the road to a Nobel Prize. Those conditions can't hurt, but they can only do so much: a large number of studies suggest a heritability for intelligence of around 50 percent; some studies put this number at greater than 80 percent.

A Swedish twin registry has information on 25,000 same-sex twins born over a period of about seventy years. Nancy Pedersen and her colleagues at the Karolinska Institute in Stockholm studied about three hundred of these twin pairs, including identical twins reared apart, identical twins reared together, fraternal twins reared apart, and fraternal twins reared together. They found that intelligence scores of identical twins who grew up apart (most having been separated by the time they were two years old) correlated much more closely than the intelligence scores of fraternal twins who grew up in the same home.

Another study, the Texas Adoption Project, followed about three hundred children who were adopted within a few days of birth and grew up entirely with their adopted families. John Loehlin and his coworkers at the University of Texas at Austin looked at results of intelligence tests given to the children at age seven and at age seventeen, to their adoptive parents, and to their biological mothers. The test scores of the biological mothers correlated significantly with those of their children who had been given away for adoption seventeen years earlier, whereas there was no correlation between intelligence scores of the children and their adoptive parents. The researchers estimated that 78 percent of intelligence is inherited.

Addictive behavior, or the predilection to become addicted, is another trait whose heritability is controversial. Addiction includes physical dependence and symptoms of craving after chronic substance use, as well as behavioral dependence—the inability to stop an activity even though the consequences are severe. Alcohol, tobacco, and illicit drug use is estimated to contribute to one in eight deaths worldwide.

Yet most people who try habit-forming substances do not become addicted. One study estimates that the probability that someone who tries a substance once will become dependent on it ranges from about one in three or four for tobacco and heroin to about one in six or seven for cocaine and alcohol to about one in eleven for marijuana. The effect of genetics on the vulnerability of individuals to becoming addicted to these substances varies widely. Studies of twins suggest that persistent smoking and nicotine dependence is about 70 percent heritable, alcohol dependence is 50 to 60 percent heritable, and addiction to most other substances is 20 to 35 percent heritable. These studies also indicate that other disorders such as antisocial personality disorder and conduct disorder are often associated with addictive behavior.

Psychiatric disorders often have a large—sometimes, surprisingly large—genetic component. Schizophrenia, for which the lifetime risk is approximately 1 percent, has a heritability estimated at around 85 percent. Bipolar disorder (also known as manic-depressive illness because it is characterized by episodes of extreme elation (mania) alternating with episodes of depression) also has an individual lifetime risk of about 1 percent. Twin, sibling, and family studies all point to a strong genetic basis of bipolar disorder: a heritability of 80 to 90 percent. Depression (also known as unipolar disorder), for which we have an individual lifetime risk of around 10 to 20 percent in the United States, undoubtedly has a genetic basis: its heritability may be as high as 70 percent.

Susan Middlebrook's unusual sleep-wake cycle is obviously heritable. This syndrome has revealed something striking about circadian rhythms. Much of what's known about these rhythms was originally worked out in *Drosophila melanogaster,* the tiny fruit fly that buzzes about the bananas left on a kitchen counter. Remarkably, the fly's clockwork mechanism, the proteins it uses to reset its timing each day, works much like ours does. The first mutations to be identified that affect these rhythms, which caused flies to have either a shorter or a longer cycle, or no rhythmic cycle at all, were in a gene given the name *period.*

In 2001, Louis Ptácek, Ying-Hui Fu, and their colleagues at the University of Utah identified a human mutation that causes FASPS. It turned out to be in a gene that encodes a protein with an amino acid sequence very similar to the fly *period* protein, so the human gene was dubbed *Period2.* This was extraordinary evidence of gene conservation over hundreds of millions of years of evolution. Even stronger evidence came from the finding that the single amino acid change in the Period2 protein that advanced Susan Middlebrook's clock corresponds to a mutation in fruit flies that advances the clock of that simple organism.

What about the heritability of obsessive-compulsive disorder, the illness that struck Jean and Jill? This disease, affecting about 2 percent of the population, has remarkably diverse symptoms but generally includes four major ones: obsessions and checking behavior; a need for symmetry and order; excessive washing; and hoarding tendencies. Individuals differ in their age of onset, the duration of the illness, and the types of symptoms they display. Many sufferers have tics, and some also have depression, phobias, separation anxiety, or disruptive behavior. Some are troubled by harmful sexual or religious obsessions, or by trichotillomania, compulsive hair pulling. Others may exhibit grooming behaviors.

This diversity of symptoms suggests that obsessive-compulsive disorder may exist in different forms that have different genetic causes—in other words, there may be multiple genetic routes to a group of disorders that have all been given a single name. So it is not surprising that twin studies almost always point to a substantial heritable component for this illness, ranging from 25 percent to 80 percent. Thus, the most probable explanation for Jean and Jill's similar illnesses at similar ages is the young women's genetic similarity, although other factors may have contributed to their condition.

Autism is a heart-breaking diagnosis for a parent to receive, and so it was for Lewis's parents. The prevalence of autism went up more than fivefold in the 1990s, and now may be as high as 1 percent of children. Much of the increase may be due to broader diagnostic criteria and increased physician and parent awareness.

Numerous studies indicate that genetic factors are the main cause of autism. For example, siblings of an autistic child have about ten times the risk of having the syndrome as the general population. A twin study in the United Kingdom yielded a heritability estimate of more than 90 percent for a broad set of autistic symptoms. Many genes are suspected of being involved in the syndrome. They haven't been identified yet, but when they are, the diagnosis of autism, which can be difficult and often confusing, will be easier and more precise. Accurate and early diagnosis will also enable earlier intervention, which will greatly improve the prognosis for these children.

Estimating the genetic component of human behaviors and psychiatric disorders may help to remove a lingering stigma attached to people with mental illness—a misplaced sense that these are character flaws—and it may inspire more people to seek treatment. In addition, these studies all point to environmental components—generally still to be teased out—that interact with the genetic ones. That's good news, since we have some control over the environmental inputs to disease. It is notable that even for the most heritable illnesses, such as schizophrenia or bipolar disorder, the heritability is never 100 percent, and even identical twins are never 100 percent concordant. Thus, even when genetics has a strong effect, it is not absolutely deterministic, so hope should never be abandoned. Finally, identification of the gene variants contributing to a disorder can assist and sharpen diagnosis, which will lead to earlier and possibly more effective treatments.

A high heritability does not imply that just a single gene is involved. In fact, for most of these disorders it is clear that many genes are involved, and no single gene is likely to explain most of the variability. By establishing a genetic basis for these disorders and identifying families that carry the causative changes in their DNA, geneticists can pinpoint the genes that are responsible.

Identification of such genes often leads to insight into the disease that can be tested in organisms suitable for experimentation, such as the fruit fly or the mouse. Analysis of those genes is sure to illuminate disease mechanisms. Just knowing what the relevant genes are will allow individuals to be tested for the gene variants that put them at risk of the disease, and, should they carry some of those variants in their personal DNA code, make them or their parents vigilant about early signs of the disease. And the genes and proteins that are implicated provide potential targets for new drugs that promise to improve the lives of people like Susan, Jean and Jill, Lewis, and many of the rest of us.

III Finding the Gene

11 Mistakes Happen: The Mutations of Cancer

The Gross sisters, Pauline and Tilly, were well known in Ann Arbor, Michigan, for their skill as seamstresses. The sisters served an upscale clientele in Gay Nineties Ann Arbor, including Dr. Aldred Scott Warthin. A renaissance man—accomplished musician, certified music teacher, scholar and scientist, author of three books—Warthin was a pathologist on the faculty at the University of Michigan, a position he held for thirty-six years.

Warthin admired Pauline Gross's work and employed her to supplement his wardrobe. He also recognized her intelligence and enjoyed his conversations with her. It was during one of those conversations, in 1895, that Warthin learned of Pauline Gross's awful premonition: she would die young. "I'm healthy now, but I fully expect to die an early death from cancer. Most of my relatives are sick, and many in my family have already passed on." Her great-grandfather, who had emigrated to Michigan from Plattenhardt, Germany, in 1831, died of intestinal cancer at the age of sixty. Six of his ten children died of cancer, as did thirty-three of his seventy grandchildren. And Pauline, just as she had feared, succumbed to the disease before her thirtieth birthday.

Why was cancer so prevalent in this family? His interest piqued, Warthin investigated the family medical history. He confirmed Pauline's contention that cancer was passed down through the generations, and concluded that "in certain families [there is] an inherited susceptibility to cancer." For thirty years he studied the disease that had struck down Pauline and so many of her relatives. His work earned Dr. Warthin the title "father of cancer genetics"—he was the first person to document the genetic basis of cancer.

Warthin understood that Pauline's family harbored a genetic defect—a mutation, though he did not know what that is and did not use that word—that caused those who inherited it to develop cancer. That he reached this conclusion is all the more remarkable because he did so before anyone knew anything about genes and Mendel's rules of inheritance: when Warthin began his studies, Mendel's work was yet to be uncovered, and the discovery that genes are made of DNA lay fifty years in the future.

Mutations are bad: they can lead to cancer. But they also provide the spice of life. Stroll down a major thoroughfare of any metropolis in America and you'll see a remarkable diversity of people. People with skin of the darkest ebony, the lightest ivory, and every shade in between. People with eyes of blue, of brown, and of all colors in between. People with full lips, thin lips, and all thicknesses in between.

There are approximately six and a half billion people on this planet, and if we examined them all in one big police lineup we could tell every one of them apart (except for the identical twins, and even there a sharp eye could distinguish many of them). And the child that is being born at the very moment you are reading this will grow up to look like no other person on earth. That is why the gold standard at the airport security gate is still the photo ID. Each person's appearance is truly unique.

Despite this individuality, if we determined the personal DNA codes of several of the people in that global police lineup (not quite feasible yet, but it will be soon), we would find that they are remarkably similar: The DNA codes of any two people are about 99.9 percent identical (ignoring for the moment the little Y chromosome present only in males). For every 1000 base-pairs of DNA, only a single one (on average) will be different between two individuals. That's true for *any* two individuals on Earth, regardless of what continent they hail from and what racial or ethnic group they belong to. (There will, in addition, be many small and large insertions and deletions of base-pairs.)

How can such little difference in our personal DNA codes lead to such an enormous diversity of physical characteristics? This is one of the major unanswered questions in biology, and the fact that some of those DNA sequence differences influence your susceptibility to disorders such as cancer, diabetes, and heart disease only raises the stakes.

Perhaps our high degree of DNA sequence identity should be unsurprising. After all, each of us has the same body plan, with internal organs forming in the same place, arteries and veins and nerves traversing the same territory, and sets of bones, muscles, cartilage, and tendons arranged the same way to provide structural support. We all have roughly equivalent dimensions: none of us grows to be the size of an elephant or stops growing after reaching the size of a mouse. At a behavioral level, we all learn a language, recognize our family members, and experience a range of emotions that are undeniably human.

On the other hand, maybe our DNA sequences are not so similar. With a genome of three billion base-pairs in each set of chromosomes, that 0.1 percent difference means that every human has about six million differences in his or her personal DNA code from anyone else in the global police lineup. Six million is more than the number of DNA base-pairs in the blueprints for entire (albeit one-celled) organisms, so maybe we should be wondering why we aren't even more different from one another than we are. Why don't some of these changes in our DNA sequence cause some of us to have nine kidneys, or to develop in only nine weeks of gestation, or to grow to be nine feet tall?

The reason we all look roughly similar is that our twenty thousand or so genes must coordinate precisely with each other to enable a fertilized egg to become first an embryo, then a fetus, and finally a baby (as discussed in chapter 4). Any changes in the DNA sequence that resulted in a set of instructions that lead to anomalous formation of organs—too few, too many, too different—would be fatally incompatible with the requirements of the developing embryo. In fact, developmental problems like these happen surprisingly frequently: up to 50 percent of all conceptions produce a defective embryo that is spontaneously aborted early in a pregnancy (often without the mother's knowledge), and at least half of those cases are likely due to changes (mutations) that occurred in the DNA of the fetus.

The constraints on the human form make it even more remarkable that we can distinguish *everyone* in the global lineup. The range of subtle variation that exists is astonishing: ears that stick out a bit more than average, eyes painted from an enormous palette of colors, lips that turn up, turn down or turn crooked. Where does all that variation come from? It comes from mutations, which are mistakes made in copying DNA. Every sperm and egg contains one copy of a person's complete set of genes. Although it's a quite good copy, it's not perfect: perhaps 200 new mistakes—differences in the DNA base-pairs from the parent's template DNA—arise in each set of chromosomes when they are copied and passed on to the next generation. That's remarkable accuracy for a biological machine composed of at least twenty parts that has to copy six billion base-pairs of DNA, but these mistakes accumulate as they get passed down through the generations.

A few million mutations, give or take some, accumulated in the generations from early humans leading up to us. Most have no effect on our appearance or health or anything else, but some of them, perhaps fewer than one hundred thousand in each person, are responsible, in their almost limitless combinations, for the astounding diversity of the global police lineup. Some are in genes that encode enzymes that work to synthesize the pigment of our skin. A mutation that enhances the function of one of those enzymes might cause more pigment to be synthesized, resulting in skin of deeper ebony; a mutation that cripples the function of one of the enzymes might cause less pigment to be synthesized, resulting in skin of lighter ivory.

Some DNA sequence variations are in genes that encode proteins that determine the pigmentation of the eye. Some cause one of those proteins to be more active and make more blue pigment; others diminish the function of the protein and lead to darker-colored eyes. Some of the one hundred thousand or so DNA sequence changes are in genes that encode proteins involved in the development of facial features during gestation of the fetus. They might result in full lips by causing more of a critical growth factor to be produced, resulting in more cells being recruited for development of the lips. Or they might cause less of the growth factor to be made, resulting in thinner lips. If the copying of DNA were more accurate than it is, the global lineup would be much more monotonous.

Now let's line up all one hundred billion of our colon cells and examine them. Unlike individuals in the global police lineup, all these cells look the same. Well, of course they do: They're clones, all descended from one cell that, early during life in the womb, committed itself to give rise to the colon. And that one cell was derived from a single cell produced at conception by the union of one of Dad's sperm and one of Mom's eggs.

That first colon cell and its descendants divided thousands of times in the course of forming and continually replacing the cells of your colon—almost as many generations as modern humans underwent in their entire history on earth. But in the copying of DNA that occurred for each of these divisions, mistakes were made. It was unavoidable. Not very many mistakes, and most were of no consequence, but every once in a while a mistake was made in an important gene.

If a mistake inactivated a gene that is required for the cell to stay alive, no harm was done: one colon cell will not be missed with billions of others just like it. But if a mistake inactivated a gene that regulates growth—one that applies the brakes to a mature colon cell, whose days of cell division should be over—the cell starts growing again, and can eventually become a mass of cells called a tumor.

If we line up the cells of that colon tumor and examine them we see something quite startling: freaks of nature! They might remind us of the poor denizens of the circus tent: the two-headed lady, the hairy man, the seven-foot giant, and the three-foot dwarf. Even if these are the first cells you have ever seen, you will likely have little trouble telling them from their normal siblings. They are ugly. But worse than just being ugly, these cells don't do their colonic jobs. They do just one thing, and they do it all too well: they divide and divide and divide some more. And when they split off from the colon and migrate to distant sites in the body, they continue to divide. Eventually they overwhelm the body. Sadly, they overwhelmed Jay Monahan, the husband of the CBS News anchor, Katie Couric. Monahan was forty-one years old when his cancer was discovered, and he succumbed to the disease nine months later, leaving behind a shocked and grieving wife and two young daughters.

A few months after his death, Katie Couric told an interviewer, "Frankly, I still can't believe it. When I think about it, it just permeates every cell in my body. You can forget about it temporarily. But then the grief comes like a huge wave and like a horrible invasion of your heart and soul."

Monahan's death was especially difficult for Couric because she never saw it coming. "He had no symptoms. Jay was tired, but we thought it was because he was flying back and forth to California for his job. And we had two small children bringing every bug in America into the house. We had no idea what was going on in his body. This cancer is so insidious."

Had Monahan had a colonoscopy exam a few years earlier, the cancer might have been detected when it was treatable, when it was not yet growing vigorously and hadn't spread past the colon. "My husband's mom, who is now in her sixties, has ovarian cancer—it was diagnosed about three years ago," said Couric. "His grandmother died of breast cancer. There is a growing body of thought that these glandular cancers—breast, ovarian, uterine, prostate, and colon—may be related, that there is, perhaps, a genetic link. If we had known this, Jay might have been screened for colon cancer."

Two years after her husband's death, in March 2000, Couric underwent a colonoscopy live on national television to raise awareness of the disease and proclaim the benefits of the exam. "I want everyone to understand that women as well as men get colon cancer," she said. "Young people as well as old people get it. And frankly, very few people take advantage of the preventive tools that exist. Too many Americans don't get tested because they don't want to talk about that part of their body. . . . I think we have to use the words, say them. Colon. Rectum. Bowels. The more matter-of-fact you are with the language, the more it helps. You can't be squeamish about it. It might cost you your life."

In 2000, Couric testified before the Senate Select Committee on Aging: "I . . . have a dream that sometime in the near future everyone could have their colonoscopies . . . and that they do it before they become symptomatic—because when symptoms start to present themselves, oftentimes the disease has already progressed." In 2004, Couric established the Jay Monahan Center for Gastrointestinal Health at the New York–Presbyterian/Weill Cornell Medical Center, which offers screening and treatment for the disease. Katie Couric is doing all she can to prevent others from suffering the kind of grief she experienced when she lost her husband to colon cancer. And it's having an effect: after her public colonoscopy there was a 20 percent increase in colonoscopies nationwide.

What drove Jay Monahan's and Pauline Gross's cancers were mutations—mistakes in copying the DNA. Some mutations make the cells divide when they should not, causing them to grow out of control; some let rogue cells escape detection by the immune system when it tries to hunt them down and eliminate them; and some enable cells to find clever new ways to commandeer nutrients and oxygen when the body tries to protect itself by starving them or choking off their air supply. These cells accumulate mutations relentlessly, becoming increasingly aggressive, taking over more and more of the body. They are like zombies that keep coming back no matter how often they are whacked over the head with a shovel.

In the late stages of cancer there can be up to 150,000 mutations in the cells of the tumor—150,000 DNA sequence differences from the innocent colon cell that gave rise to the tumor. So although a few copying mistakes in the chromosomes of the sperm and egg are generally viewed as good (variety is the spice of life), they are decidedly not so good when they drive inappropriate division of other cells of your body.

Many kinds of copying mistakes get made in the DNA copying process. The simplest type is a substitution of one base for another. Most of the time the copying machinery puts the correct base in the copy—an A opposite a T and a G opposite a C in the template—but every once in a while it messes up and inserts the wrong base: say, a G instead of an A opposite a T, or an A instead of a G opposite a C.

Recall that triplets of the bases specify amino acids according to the genetic code. Here are five triplets in the middle of a gene, with the amino acid each triplet specifies shown below:

... CTG CAG **TTG** GAG AGC ...

leucine glutamine leucine glutamate serine

A mutation in a person that changes the first base of the TTG, leucine, triplet to a G would change the base sequence to become:

... CTG CAG GTG GAG AGC ...

Because GTG specifies the amino acid valine, this gene would now specify a protein with these amino acids:

... CTG CAG **GTG** GAG AGC ...

leucine glutamine valine glutamate serine

In most proteins, which contain hundreds of amino acids, the change of this one amino acid may have little or no effect. Or it may disrupt the structure of the protein enough to cause a complete loss of function of the protein. Mostly, though, single amino acid changes have little effect on the chemical properties of the protein. And some single base substitutions in DNA don't change the amino acid sequence of the protein at all, because the genetic code has multiple different base triplets that specify the same amino acid.

Insertions and deletions of base-pairs, by contrast, almost always have a large effect on the encoded protein. These mutations typically cause all of the amino acids in the protein to change from the point of the insertion or deletion onward to the end of the protein, because they shift the "register" of the triplets, so that all the bases form new groups of three. Consider the insertion of a G base after the fourth base of the DNA sequence:

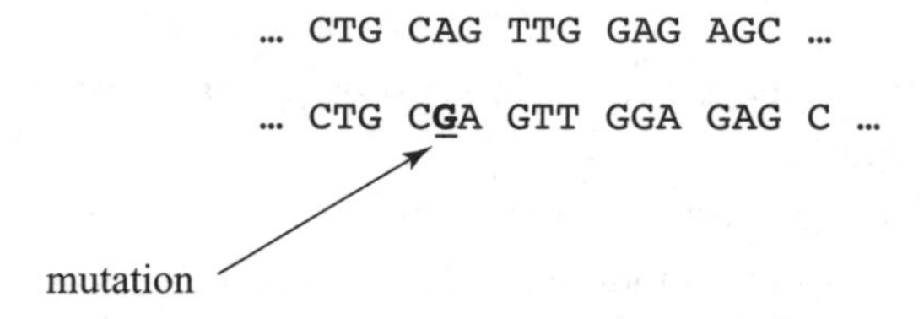

This mutation creates a new triplet CGA, and changes as well all subsequent triplets. Compare the old sequence without the mutation with the new sequence, containing the mutation that adds a base, to see what happens to the protein:

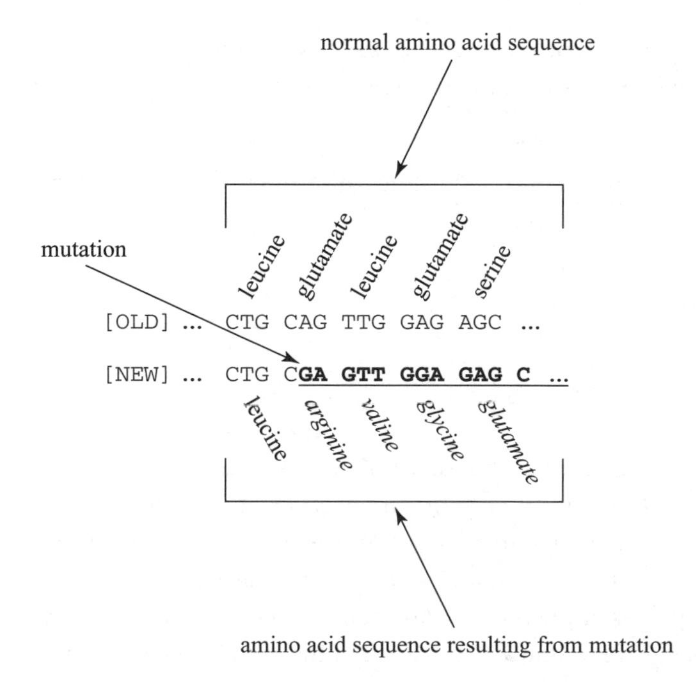

Because insertions and deletions change many amino acids, not just one, they nearly always result in a defective protein that cannot do its job.

Insertion and deletion mutations can span many base-pairs. Indeed, millions of base-pairs can be inserted into or deleted from a chromosome. Whole genes, even stretches of DNA that encode many genes, can be deleted, or duplicated to result in extra copies of a gene, which leads to increased, sometime toxic, levels of its encoded protein.

All of these kinds of changes, and more, that occur during the production of sperm and eggs are responsible for the diversity of the global human lineup. But when they occur in other cells of our bodies, they can lead to the kinds of cancers that afflicted Jay Monahan and so many members of Pauline Gross's family.

It took almost a century after Pauline Gross confided her fear of cancer to Dr. Aldred Warthin for a defect in the gene responsible for the disease in her family to be identified. The gene, called *MSH2*, encodes a protein that is part of the quality-control machinery that fixes copying mistakes in DNA.

One of the reasons copying of the genome is so accurate, with only a few mistakes in every billion base-pairs copied, is that the cell has several ways to repair copying errors in DNA. One of them, called mismatch repair, finds and fixes misincorporated bases that occur when the copying machinery inserts the wrong base—for example, an A in the new copy opposite a C in the template instead of the G that should go there. Those two bases, the A and the C, are mismatched—they are improperly paired, like a two-prong 110-volt plug trying to fit into a three-prong 220-volt socket—and the mismatch repair machinery recognizes that and fixes it.

A misincorporated base happens perhaps once in every five million to ten million base-pairs copied. Nearly all—99.9 percent–of the mismatches get fixed by the highly efficient mismatch repair machinery. A few of them—about one in a thousand—slip by and become the mutations we have been talking about.

But if you are unfortunate enough to inherit in your personal DNA code a defect in one of the genes specifying a protein of the mismatch repair machinery itself—as Pauline Gross was—then all of your cells carry that defective gene. This mutation has no immediate effect, because there's another, good copy of the gene on the other chromosome (recall that there are two copies of each gene, one on the chromosome from Mom, and one on the chromosome from Dad), and this good copy provides a functional version of the protein necessary for mismatch repair. But if one of the random mutations that occurs in our cells as they divide throughout our lifetime happens to affect the other, good, copy of the mismatch repair gene, then that cell loses its ability to fix mismatched bases.

The consequence will be a greatly increased rate of mutation in that cell—one thousand times the normal rate. As that cell divides it will accumulate mutations at a furious pace. Some of these are bound to eventually affect genes involved in controlling growth of the cell, so cancer is almost inevitable: About 90 percent of people who inherit a defective *MSH2* gene get cancer.

People who inherit this defective gene begin to develop cancer in their twenties, because that's how long it takes for the other, good copy of the gene to acquire a mutation in one of the cells, and then for the cells with that mutation to grow and accumulate the further mutations in other genes that lead inexorably to the tumor that sends the unfortunate person to a clinic. Most of them get cancer by the time they are in their forties; Pauline Gross died before she saw thirty. The cancer usually occurs in tissues with rapidly dividing cells, such as the colon and the endometrium, the lining of the uterus, because each cell division provides another chance to pick up more mutations. Pauline Gross died of endometrial cancer.

Now that the causative gene has been identified, Pauline's relatives can find out whether or not they're at high risk of developing cancer. Her great-grand-niece, Ami McKay, a thirty-eight-year-old Canadian author, learned that she had the defective gene. "Peering into my DNA did indeed change me," she said. "As you might imagine, it caused me to take an immediate inventory of my health . . . I became vigilant about making doctor's appointments and setting up annual screenings. But the results also infused my life with a curious sort of fearlessness. . . . Whenever I think of hesitation, of saving my imagination for another day, I think of Pauline."

Her sister Lori learned she did not have the defective gene. Forty members of the family were tested for the defective gene after giving blood at a family reunion in Ann Arbor, Michigan, the Gross sisters' hometown; five of them received unwanted news. More happily, a total of ninety-seven family members were told they do not carry the defective gene, either because of the results of the test, or because their parents or grandparents were known not to carry it. But they should not let down their guard: remember, a history of cancer was not so obvious in the family of Jay Monahan, Katie Couric's husband.

Mutations are egalitarian: they are happening in all of us all the time. If we live long enough, it's likely they'll eventually get us. But in the meantime, enjoy the diversity they provide in the global police lineup.

12 Reshuffling the Genetic Deck: A Cancer Gene in the Neighborhood

Seymour Benzer was the scientific equivalent of Bo Jackson, the star athlete who dodged defenders on football fields for the Los Angeles Raiders and hit towering home runs on baseball diamonds for the Kansas City Royals, the Chicago White Sox and the California Angels. Benzer, active as a professor at the California Institute of Technology in Pasadena until his death in 2007 at age 86, was trained as a physicist but made breakthrough discoveries in two unrelated areas of biology. One of those discoveries is central to our story.

Benzer obtained his doctoral degree in physics from Purdue University shortly after the Second World War, and stayed on there to work in solid-state physics. But even as a young faculty member Benzer knew he wanted to move into biology. Like many physicists of that day, he had read *What Is Life?*, the influential 1944 book by the Nobel Prize–winning physicist Erwin Schrödinger that challenged postwar scientists to seek an answer to the question posed by its title. The book inspired Benzer to attend a course in the summer of 1948 at the Cold Spring Harbor Laboratory on Long Island to learn about the genetics of bacterial viruses. These tiny beasts are called bacteriophages, or "phages" (from the Greek "to eat"), because they live on—actually eat—bacteria.

Benzer studied phages for nearly two decades, making a remarkable finding about DNA in the process. In the mid-1960s, Benzer shifted his focus to neurobiology, using the fruit fly *Drosophila melanogaster* to identify genes that control behavior. He continued those studies for forty years, pioneering a whole new field of study in the process. He was a remarkable scientist.

Benzer loved food, and was adventuresome in his culinary choices. As a visitor from 1951 to 1952 at the Pasteur Institute in Paris, he would bring for lunch such delicacies as South African caterpillars, cow's udder, bull's testicles, or filet of snake. One day in Paris his young daughter woke up with her eyes swollen shut. When their doctor asked whether she had eaten anything unusual lately, Benzer was reluctant to reveal the truth.

There's no accounting for taste, you might say. But actually, there is. People taste things differently. For example, phenylthiocarbamide (PTC) is a sulfur-containing compound similar to chemicals found in leafy green vegetables such as cabbage and broccoli. Chew on a small strip of paper impregnated with PTC and it's likely you'll spit it out; it tastes extremely bitter. But about 30 percent of Americans sucking on such a strip will look at their fellow tasters who are grimacing and ask, "What's all the fuss?" They can't taste a thing.

This simple taste test, given every year to thousands of high school students in biology classes, demonstrates simple Mendelian inheritance: a single human gene controls whether or not we can taste PTC. This gene, which goes by the name of *TAS2R38*, encodes a protein found in the tongue that is a taste receptor. The functional version of the gene that 70 percent of us have, which enables us to taste PTC, is dominant over the nonfunctional, non-tasting version.

The *TAS2R38* gene resides at a position in the genome designated 7q35, meaning in region 35 of the long ("q") arm of chromosome 7. (The short arm of a chromosome is called p, for "petite"; the arms are joined at a "neck" that is visible under the microscope.) A more precise description of the position of the *TAS2R38* gene is that it begins with the 141,318,900th base of chromosome 7. Keep in mind that the version of *TAS2R38* you have doesn't much affect the quality of your life.

Now let's walk down chromosome 7 to a gene named *MET*, located at 7q31. If you're a New Yorker, this name may conjure up opera singers, artists, or outfielders, but in fact it derives from *m*esenchymal-*e*pithelial *t*ransition factor, a substance that plays a role in forming tissues during embryonic development. The *MET* gene encodes a receptor, a protein that sits on the outside surface of cells looking to hook up with a specific molecule, in this case with another protein called hepatocyte (liver) growth factor.

This particular growth factor, one of many coursing through our bodies, is manufactured in various tissues, including the liver, kidney, lung, and brain, and is sent out to circulate, looking for its receptor. When it finds one, it latches on tightly and "tickles" the receptor, causing it to send a signal into the cell that tells the cell to grow and divide.

Unlike *TAS2R38*, the version of the *MET* gene you have *does* affect the quality of your life: some mutations in the *MET* gene cause the receptor to go into overdrive, stimulating cells to divide when they shouldn't, which leads to uncontrolled growth. If your personal DNA code contains one of these mutations, the result is hereditary papillary renal cancer, a cancer that often develops at many sites in both kidneys. Fortunately, it's not one of the more aggressive kidney cancers; if it is caught before it's too far advanced, the long-term prognosis is often good.

Because of how the altered form of *MET* promotes tumor formation, or oncogenesis (from *onco*, meaning tumor, and *genesis*, meaning birth), the gene is known as an oncogene. The normal function of many oncogenes is to regulate cell growth, often by encoding growth factors that act in signaling pathways to control whether or not cells divide. Other well-studied oncogenes operating in such pathways include those with the designations *RAS*, *SRC*, *JUN*, and *MYB*, names derived from the cancer-causing viruses in which these genes were first found.

The growth-activating pathways these oncogenes operate in would do Rube Goldberg proud: the tickling of the receptor by binding of the growth factor flips a switch that opens a valve that nudges a protein that bangs into another protein that tips over a ramp along which rolls another protein that eventually lands on a specific site of a chromosome, where it turns on a gene whose protein product steps on the accelerator of cell division. Of course, scientists don't describe it quite like that, but if you broke open a cell and peered inside, as scientists do, you would see that it is full of all kinds of convoluted contraptions not so different from Rube's machines, which keep the cell humming along when they are working but wreak havoc when they malfunction.

Why, you may be wondering, are we discussing these two genes together? What is the connection between them? The *MET* gene begins at base-pair 116,126,375 of chromosome 7—about twenty-five million base-pairs away from the *TAS2R38* gene (25,192,525, to be exact). This may seem a long way apart, but in fact, the two genes are really just around the corner from each other in our three-billion-base-pair genome (the distance between them is only about 0.8 percent of the genome).

Now we come to a key point concerning the relationship between the *MET* and the *TAS2R38* genes. These two genes are what geneticists call "linked." Here is how it works. Suppose that your mother comes from a family in which there have been many cases of hereditary papillary renal cancer, and she's already had a couple of bouts of the disease herself. It's clear she has inherited the mutant form of the *MET* gene, which we'll call MET^*; we'll call the normal (prevalent) form met^+. The MET^* mutation is dominant, so (as we discussed in chapter 7) if you inherit this form of the gene from your mother—and there's a 50 percent chance you will—you'll have to deal with this disease yourself. Is there any simple way to know which version of the gene from Mom is in your personal DNA code?

Maybe. It is easy to figure out that Mom is a PTC taster, like the majority of the population, and that Dad is not because he can chew PTC paper all day and not distinguish it from a wad of newspaper stuffed in his cheek. So Dad must have two mutant versions of the (recessive) taste receptor gene. He's never had kidney cancer, so it's unlikely that he has the rare MET^* version of that gene as Mom does, because if he did it likely would have caused the disease already. If we write the functional version of the taste receptor gene as TAS^+ and the mutant version as tas^- and use a long rectangle to represent a chromosome, then the configuration of Dad's two copies of chromosome 7 must be:

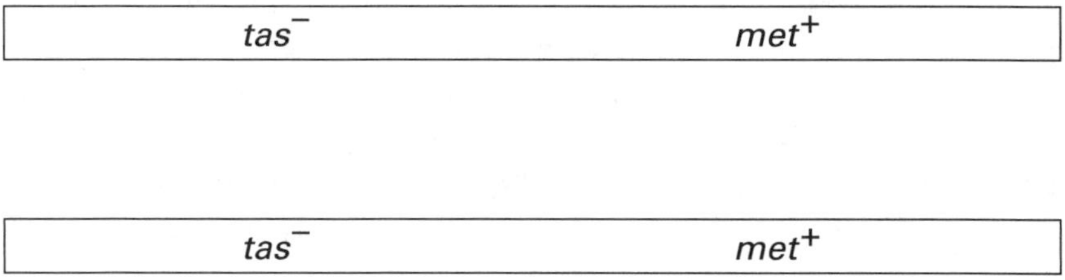

tas^- met^+

tas^- met^+

Since Mom can taste PTC, she must have at least one good version of the *TAS* gene:

TAS^+

Once you learn her mother can't taste PTC, you can deduce that Mom has *only* one good *TAS2R38* gene, because all her mother could have given her was the *tas*$^-$ version of the gene that encodes a nonfunctional taste receptor. Her functional (*TAS*$^+$) tasting gene could only have come from her father. Thus, her taste receptor genes on her two chromosomes 7 are:

TAS$^+$

tas$^-$

You need to know one more key fact: Your maternal grandfather died of renal cancer. This tells you that your mother got her cancer-causing (dominant) *MET** gene from her father, so it must be on the same chromosome 7 as the *TAS*$^+$ gene she got from him, making Mom's chromosome configuration:

TAS$^+$ *MET**

tas$^-$

Your maternal grandmother never got kidney cancer, so you can assume she had the normal (*met*$^+$) form of the *MET* gene. So Mom's two chromosomes must be:

TAS$^+$ *MET**

tas$^-$ *met*$^+$

We can now see that in your maternal grandfather, the functional (tasting) version of the *TAS2R38* gene was on the same chromosome as the cancer-prone version of the *MET* gene, and that this arrangement was passed on to your mother. Geneticists say that the TAS^+ and MET^* versions of these genes are "linked" to each other, as are the tas^- and met^+ versions of those genes. All of this is summarized here.

Maternal grandfather's chromosomes 7:

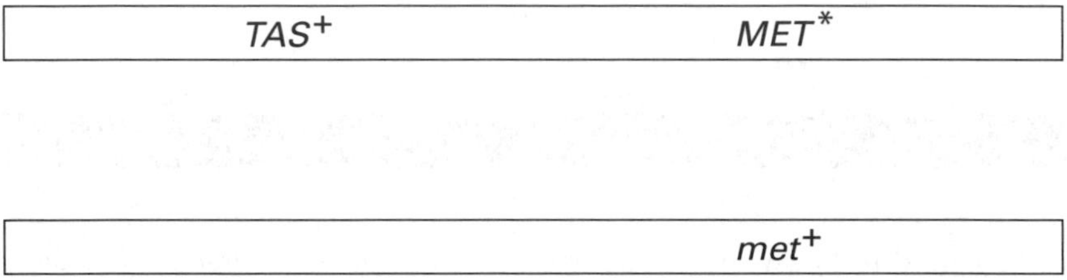

TAS^+ MET^*

met^+

Maternal grandmother's chromosomes 7:

tas^- met^+

tas^- met^+

Mom's chromosomes 7:

TAS^+ MET^*

tas^- met^+

Previously we showed Mom's chromosomes in black to distinguish them from Dad's chromosomes, shown in white. But of course Mom got one set of her chromosomes from her father and one set from her mother. So now we show the one she got from her mother in black and the one she got from her father in white.

If you've followed this reasoning you'll realize that you should be able to tell whether you inherited your mother's cancer-causing MET^* gene simply by determining whether you can taste PTC. Since the only functional *TAS2R38* gene among the four versions of this gene present in your parents is on the same chromosome as the mutant MET^* gene that brings renal cancer, you say: "If I can taste the bitter compound then I must have inherited Mom's chromosome that has the TAS^+ gene that encodes a functional *TAS2R38* receptor, and since I now realize that TAS^+ and MET^* are linked, I must also have inherited the defective MET^* gene that's on that same chromosome." The *TAS2R38* gene, which encodes a protein that has nothing to do with cancer risk, should serve as an easily identifiable marker—a bellwether gene—that tells you key information about the nearby *MET* gene.

It would be nice if it were that simple, but there's a wrinkle. We know that when sperm and egg cells form, the chromosome number decreases by half. Each of the sperm or egg cells gets a single chromosome 1, a single chromosome 2, and so on. The wrinkle is that before the two members of a pair of chromosomes go their separate ways into different sperm or egg cells, the genes they carry play a game of musical chairs, and some of the genes trade places. Genes on the chromosome in Mom that she inherited from her mother can exchange places with their counterparts on the chromosome that she inherited from her father. Both chromosomes still carry the same kinds of genes in the same order, but since your maternal grandparents had slightly different DNA sequences (about one in one thousand base-pairs were different between them), the newly constructed chromosomes have your maternal grandmother's versions of some genes, and your maternal grandfather's versions of others. The two different versions of Mom's genes end up in new combinations in each chromosome when Mom's eggs form. This seemingly strange part of the process of producing eggs and sperm contributes a huge amount to human diversity, and provides a tool that enables geneticists to locate the genes responsible for diseases.

How does this process happen? Before your parents got intimate and conceived you, the chromosomes in their cells that produce eggs or sperm got intimate, cuddling up to one another and exchanging pieces of their DNA during their game of musical chairs. But they didn't exchange their pieces willy-nilly. The chromosomes are monogamous and pair up only with their proper partners: the two copies of chromosome 1 snuggle up and pair with each other, but not with any of the other 22 chromosomes; the two copies of chromosome 2 pair up with each other, and with none of the other chromosomes, and so on.

The chromosome pairs then exchange pieces with each other, resulting in rearranged chromosomes that are composed of chunks of each chromosome. Each chromosome in your mom's eggs has stretches that came from her mother and stretches that came from her father; same thing with the chromosomes in your dad's sperm. Each chromosome in the sperm and egg cells that joined to form you is a mosaic of your grandmother's and grandfather's versions of the chromosome. So genes from your grandparents are mixed up in the grandchildren.

How does this work? Genes are arranged linearly along chromosomes because, as you know, genes are simply stretches of a DNA molecule that runs from one end of the chromosome to the other. The genes are like trinkets hanging off a charm bracelet. Each of your cells (apart from the sperm or eggs) contains twenty-three pairs of chromosomes, one of each pair from Mom and one from Dad:

1	2	3	4	5	6	7	8	9	10

1	2	3	4	5	6	7	8	9	10

The numbers designate individual genes, but of course a typical human chromosome has many hundreds of genes.

During the generation of egg and sperm cells, the two copies of each chromosome pair up, in register, such that maternal gene 1 aligns with paternal gene 1, maternal gene 2 with paternal gene 2, and all the rest also in register, for the entire length of each chromosome. All the other chromosome pairs do the same thing, apart from the X and Y chromosomes, which have very different DNA sequences and so cannot align to each other.

Once the chromosomes have paired up, the game of musical chairs starts. The DNA strands get broken at several random places along the chromosome, but they are quickly rejoined with the help of special enzymes whose job it is to repair chromosome breaks. Sometimes the breaks are simply resealed, reconstructing the original configurations of the chromosomes (all from Mom and all from Dad). But sometimes a piece of one chromosome will join to its partner from the other chromosome with which it's paired, producing two mosaic chromosomes, each with some part of Mom's chromosome and some part of Dad's chromosome. The two chromosomes are regenerated perfectly, with loss of not a single base of DNA! Pieces of Mom's and Dad's chromosomes have traded places.

Say a break occurs after the third gene. After exchanging and rejoining, the two chromosomes look like this:

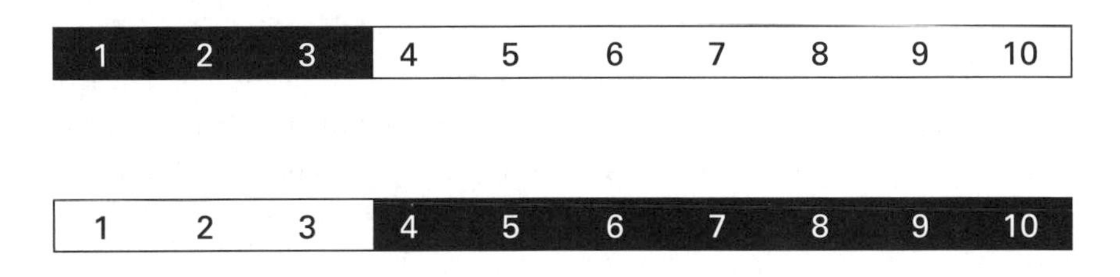

When this cell gives rise to two sperm or egg cells, one will inherit the chromosome shown here:

1 2 3 4 5 6 7 8 9 10

while the other will inherit the chromosome shown here:

1 2 3 4 5 6 7 8 9 10

Each of the pairs of chromosomes undergoes at least one of these reshuffling events, usually more, before it ends up in a sperm or egg cell.

The result of these exchanges is that genes have been reshuffled. The genes are still in the same place on the chromosome (base-pair 141,318,900 for the *MET* gene and base-pair 116,126,375 for the *TAS2R38* gene), but their versions—their DNA sequences, which are slightly different in each parent—are now in a different combination.

Before the exchange event, a certain version of a gene present in one of Mom's chromosomes that came from her mother was adjacent to a version of a neighboring gene that also came from her mother, but after the exchange that gene finds itself next to the version of the neighboring gene that was on the chromosome Mom obtained from her father. The chromosomes in you that come from your mom are thus mosaics of the chromosomes in your maternal grandfather and grandmother. Of course all of this is also the case for the chromosomes that Dad gave you—they are also a mix of his mom's and dad's genes. You can see how individual you are. You have a unique combination of different versions of each gene: your personal DNA code.

The reshuffling phenomenon was first observed and largely deciphered in the fruit fly, whose distinctive features for genetic analysis were described in chapter 4. The chromosome reshuffling in the fruit fly that could be analyzed occurred *between* genes, not *within* genes—like bracelets are rearranged by exchanges between the charms, not within the charms themselves.

In the 1950s, Benzer asked a profound question: Can the reshuffling of genetic material occur *within* a gene, as well as between genes? His question got to the heart of the nature of the gene: Is the gene divisible? This was perhaps an obvious question for a physicist, one who not so long before had been mulling over the same question concerning atoms.

You likely appreciate that if chromosome reshuffling occurs at random, the likelihood that it will occur between two specific sites only a few base-pairs apart on the DNA is exceedingly small. How small? Well, if we were trying this experiment using humans, we might look through the entire world population of six and a half billion people and not identify a particular reshuffling event. Even with fruit flies, where biologists can breed eighteen generations in a single year and look through thousands of individuals, far too few can be handled to find a reshuffling event that happens maybe once in a billion cell divisions.

Benzer's keen insight was to realize that he could answer his question about the divisibility of the gene using bacteria. Bacteria are free-living creatures that can divide every twenty minutes, yielding more than twenty-five thousand generations per year, and hundreds of millions of them can be held in a single test tube. But even smaller and more numerous creatures were available to Benzer: the viruses that attack bacteria, called bacteriophages, which, like all viruses, are little more than a tiny bit of DNA inside a simple coat of protein. Phages cannot survive and grow on their own because they lack the complex machinery to break down food and harvest the energy therein, and to synthesize their own proteins for their coat, and to carry out many other essential processes.

Using the human gut bacterium *Escherichia coli* and a gene in a phage called T4 that goes by the name *rII*, Benzer showed that reshuffling can occur anywhere within this gene. Like the atom, the gene can be split. By determining how often reshuffling happened between any two regions of the phage chromosomes, Benzer generated a map of the *rII* gene. Similar genetic maps of human mutations in different genes pinpoint the position of genes along a human chromosome. As we will see in the next two chapters, these maps allow us to find the gene responsible for a genetic disorder in a pedigree of families with the disease, or to associate variants of genes with increased susceptibility to a disease.

In the late 1950s, Benzer again went abroad, to spend a year at the Medical Research Council Laboratory of Molecular Biology in Cambridge, England, working with Francis Crick. On arriving in Cambridge, Benzer spied a book that included a promising list of local restaurants. He suggested to Crick that they have lunch at a different one each day, but Crick was unenthusiastic, preferring his sandwich and beer at the Eagle Pub. Nonetheless, Crick humored Benzer and joined him at a succession of restaurants, each worse than the last. At a place called the Firehouse, Benzer nearly choked to death on a piece of steel wool in his hash.

A month later another American scientist, George Streisinger, arrived to work at the MRC Laboratory. He came in one day and said, "Seymour, I just found out about all these wonderful restaurants in Cambridge. Let's go to a different one for lunch every day." Benzer demurred.

Let's look at the reshuffling that goes on among human genes. Because genes on the same chromosome are physically linked, literally tied to one another, if this genetic reshuffling did not occur, they would always be inherited together. Recall our two genes on chromosome 7:

TAS^{+} enables tasting of PTC

tas^{-} results in inability to taste PTC

MET^{*} leads to high risk of kidney cancer

met^{+} provides low risk of kidney cancer

Although Mom has one of each version of the two genes, she is able to taste PTC and got renal cancer because those are the dominant traits. With no reshuffling, all of Mom's eggs carry one of these chromosomes:

TAS^{+} MET^{*}

or

tas^{-} met^{+}

Dad has no increased renal cancer risk and can't taste PTC, the result of having the recessive version of both genes, so all of his sperm must contain this chromosome 7:

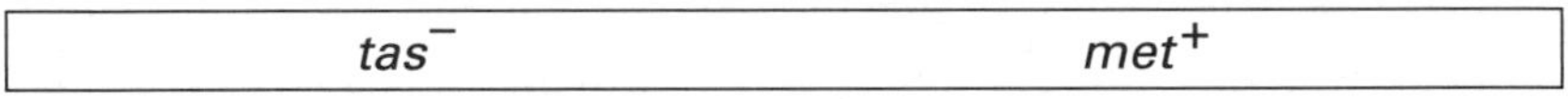

tas^{-} met^{+}

When any of his sperm are joined with one of Mom's eggs containing an unshuffled chromosome 7, the offspring will have either:

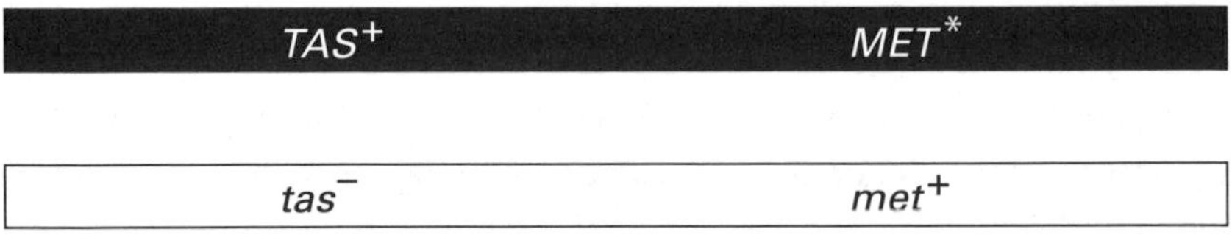

TAS^{+} MET^{*}

tas^{-} met^{+}

or

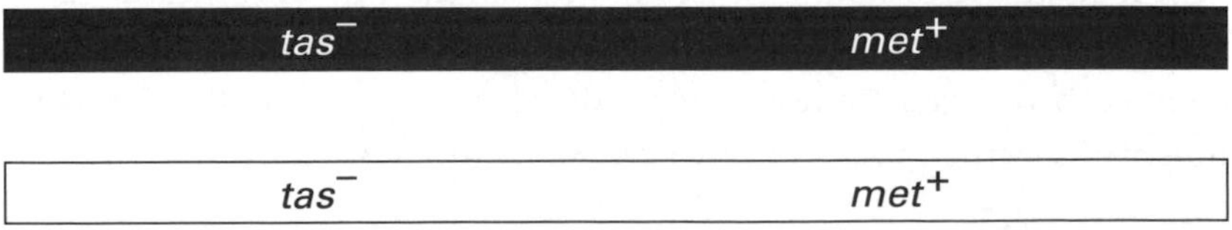

tas^{-} met^{+}

tas^{-} met^{+}

If the egg you developed from had the first combination of chromosomes, you, like your mother, will be able to taste PTC and will likely get kidney cancer. But if you developed from an egg with the second combination of chromosomes, you will not be able to taste PTC and will have no increased cancer risk, like your father. That is, without genetic reshuffling, all of your parents' children will have one or the other parental combination of genes, and the ability to taste PTC would indeed indicate an increased risk of renal cancer.

But what if these two genes reshuffled (breaking and rejoining at the "X") en route to one of Mom's egg cells:

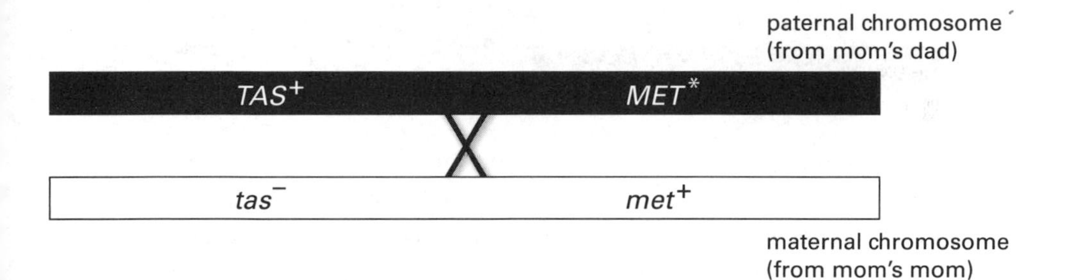

The result is new combinations of gene variants on chromosome 7:

In some of the eggs Mom produces, the *TAS*⁺ version of the gene, which your mom got from her father, is now on the same chromosome as the *met*⁺ version, which she inherited from her mother. In other eggs, the *tas*⁻ version, which came from her mother, is on the same chromosome as the *MET** version, which came from her father. So it could be that the fertilized egg that developed into you carried Mom's newly generated chromosome 7:

tas⁻	*MET**

(Of course, you will still receive a chromosome from Dad with the linked *tas*⁻ and *met*⁺ genes, since that is the only kind he has.) In that case, you'll have a high chance of developing renal cancer, but you won't be able to taste PTC, in which case you resemble neither Mom nor Dad. You would be a mixture of their two types as a result of the exchange between Mom's pair of chromosomes 7 in her maturing eggs.

Your sister, by contrast, may have developed from an egg that had the other kind of reshuffled chromosome 7 from Mom:

TAS⁺	*met*⁺

Again, your sister can only get a chromosome from Dad with the *tas*⁻ and *met*⁺ genes, so she will be able to taste PTC but have a normal low risk for kidney cancer, which also is different from either of your parents. She, too, is a mixture of your parents' types due to the exchange of Mom's copies of chromosome 7 in her maturing eggs. Since chromosome reshuffling occurs frequently, the ability to taste PTC is not a guaranteed indicator of the version of the *MET* gene you inherited.

The key point to grasp about genetic reshuffling is that the closer two genes are to each other on a chromosome—the more tightly linked they are—the less likely they are to be reshuffled and the more likely they will stay together on the same chromosome. This is because genes that are close to each other on a chromosome are separated by few base-pairs; since the chromosomes exchange their pieces at breaks between the base-pairs, there are fewer opportunities for a break and exchange between genes that are close than between genes that are far away from each other. Conversely, the farther apart two genes are, the more likely there will be a break between them at which they can reshuffle.

Returning to the charm bracelet analogy, imagine that you close your eyes and whack at the chain with a hatchet, splitting it randomly at a single location. Let's label the *TAS* and *MET* genes on the maternal set as if these two genes were closely spaced:

TAS MET

The likelihood that that the hatchet will fall precisely between the two genes is low, because this region occupies only a small portion of all the places the hatchet might land. But if the two genes lie far apart, then the maternal chromosomes would look like:

TAS MET

and the likelihood is high that the hatchet will land somewhere between the two genes.

The order and distance of genes along a human chromosome is a genetic map, and genes are defined as being one map unit apart if they are reshuffled in 1 percent of the sperm and eggs. The *MET* and *TAS2R38* genes are about twenty-five map units apart, which means that the chromosomes in about one out of four sperms and eggs will have undergone a reshuffling event between the two genes. So you could very well taste PTC yet have the good version of the *MET* gene because you inherited from Mom a chromosome 7 that contains those two versions of the genes, even though she didn't have such a chromosome 7 herself (until she produced it while making the egg that turned into you).

How were the first genetic maps generated? It certainly wasn't by studying human chromosomes, because until recently it has been difficult to map human genes: we can't do mating experiments on humans, our generation time is far too long (only about five generations per century), and our progeny are far too few for effective genetic analysis.

The first genetic maps were of the chromosomes of simple organisms, such as the bacterium *E. coli*, which lives in your gut; its parasitic phage T4, which Benzer studied; the yeast *Saccharomyces cerevisiae*, which was used to make the wine you had at dinner; and the fruit fly *Drosophila melanogaster*, which so annoys you when it swarms around your fruit bowl. The first three of those organisms have generation times measured in minutes or hours, so billions of organisms in many generations can be examined in a single experiment. Fruit flies produce a large number of offspring in just a few weeks, so flies, with a plethora of different visible traits such as deformed wings, defective eyes, and many, many more have been mated by geneticists since the turn of last century, and their resulting offspring have been examined to find traits that are co-inherited. In this way a detailed genetic map was established for each chromosome of the fly.

Biologists studying bacteria, yeast, and fruit flies weren't necessarily thinking about what happens in humans when sperm and egg cells are formed; fifty years ago it wasn't nearly as apparent as it is today how similar are the basic life processes in all creatures. These biologists were simply curious—they wanted to know how phage and flies and yeast work, and carried out the basic research that they hoped would give them the answers.

Now that we know that bacteria, yeast, fruit flies, and even the lowly bacteriophage reshuffle their genes just as humans do, it's clear that this basic research is not only worthwhile, it's absolutely essential if we are to make progress in finding cures for human diseases. In fact, proteins in humans involved in the reshuffling of chromosomes, as well as those necessary for repairing and copying DNA, have sequences of amino acids similar—in some cases extremely similar—to their counterparts in bacteria, yeast, fruit flies, and bacteriophage. And mutations in many of those genes can lead to an increased rate of cancer, just as with mutations in the *MET* gene. So the next time a young biologist tells you she's studying "DNA repair in *Drosophila*," or "the mutation rate in *E. coli*," or "DNA recombination in yeast," please don't shake your head at this ostensible waste of your tax dollars. Instead, thank her for the curiosity that may someday keep you or one of your children alive after a diagnosis of cancer.

13 A Family Affair: Mapping a Gene for ALS

Only two years after proclaiming himself "the luckiest man alive" at his tearful retirement from baseball at Yankee Stadium in 1939, Lou Gehrig succumbed to amyotrophic lateral sclerosis (ALS). Lou Gehrig's disease usually strikes in middle age, as it did Gehrig: the diagnosis of ALS that turned out to be a death sentence was handed to him on his thirty-sixth birthday. This was about a month after the first baseman took himself out of a game against the Detroit Tigers because he was no longer able to perform at the level he, his teammates, and fans had come to expect. That game ended his remarkable streak of 2,130 consecutive games played, a record that would stand for fifty-six years.

Gehrig could no longer perform at his accustomed level because he was losing control over his muscles. ALS causes its victims to slowly waste away: the nerves are increasingly unable to send signals to the muscles, first causing clumsiness but eventually progressing to near paralysis. Muscles far from the spinal cord are the first to be affected. A wave of paralysis then marches relentlessly up the arms and legs until it reaches the diaphragm, ultimately causing the victim to be unable to breathe, usually three to five years after the disease is diagnosed. Lou Gehrig's disease is indeed an awful one, for which there is no effective treatment and no cure.

Andrew Mattingly Jackson was somewhat luckier than Lou Gehrig, if you can call someone who inherits a neuromuscular disease lucky, because he contracted a rare juvenile form of ALS known as type 4 ALS, which is much milder than Lou Gehrig's disease, though still serious. Jackson was a descendant of Thomas Mattingly, one of the first British colonists of the New World and the founder of a Maryland-based clan, more than seventy of whose members have suffered from this type of ALS. Jackson began wondering in his early teens why he couldn't run very fast and wasn't more of an athlete, but he didn't think much more of it because in 1945, at eighteen, he was fit enough to pass a physical and join the U.S. Navy, and he made it through boot camp. But he couldn't ignore his diminished physical abilities for long, because one day two years into his service he found himself unable to climb the ladders onto his ship after a day of strenuous work. Like Lou Gehrig, Jackson took himself off duty, and was quickly diagnosed with ALS.

But unlike Lou Gehrig, Jackson continued to function as well as most people, and needed only a cane to get around until he was well into his fifties, when, as his strength waned, he started to use a walker. In fact, his disease was so mild that at one point physicians at Johns Hopkins University reversed his ALS diagnosis and told him that he had the much milder Charcot-Marie-Tooth disease. People with ALS4 can look forward to an increasingly difficult, but not foreshortened, life. Jackson worked for Black and Decker for forty years and, like Gehrig, retired only when he felt he could no longer perform at the highest level. At seventy-six, Jackson could still stand, albeit with the aid of a power wheelchair; could still type on his computer, albeit with the aid of a device to support his arm; and was still actively involved in his church. Unlike Lou Gehrig, Jackson inherited his disease.

Jackson traced his disease back to his great-great-great-great-great-grandparents Thomas and Elizabeth Mattingly, who left England and landed on the shores of Maryland in 1634. Jackson took great interest in his disease, and compiled an impressively complete pedigree of his clan—a record of his ancestors—spanning 350 years and eleven generations. A small section of the Mattingly-Jackson pedigree (see figure) shows three generations of the Maryland branch of the Mattingly family. The branch of the family that inhabits the western states—including, coincidentally, another great Yankee first baseman, Don Mattingly—is not afflicted with ALS. Perhaps when those Mattinglys trekked westward, the ones with the disease were unable to undertake or survive the strenuous journey.

To make sense of the Mattingly family pedigree we need to tell you the conventions used by geneticists: circles indicate females; squares indicate males; a horizontal line connecting a circle and square indicates a mating; the children that resulted appear at the ends of the vertical lines below the mating, from left to right in the order of their birth; a diagonal slash

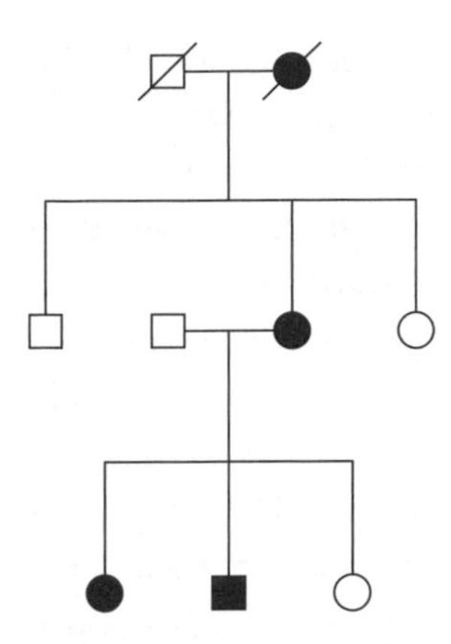

through a symbol indicates someone no longer alive. Most important: Filled-in squares and circles indicate individuals with the disease.

What pattern of inheritance does this type of ALS follow? The mutation must be dominant, because those who have the disease pass it on to approximately half of their children, and only a single parent must have the mutation to result in children with the disease. The mutation is not on one of the sex chromosomes (X and Y) because the disease afflicts both males and females. Family members alive today continue to pass on to their children the mutation that Thomas or Elizabeth Mattingly brought with them to the Maryland shore three centuries ago.

Phillip Chance, a researcher at the University of Pennsylvania, learned of Jackson's extensive pedigree and realized he could use it to identify the defective gene that Thomas or Elizabeth Mattingly passed on to so many of their descendants. The inheritance of this defective gene is called Mendelian, because it follows Mendel's two simple laws of inheritance: (1) we have two copies of each gene (though the word "gene" wasn't coined until 25 years after Mendel's death), because our chromosomes come in pairs; (2) the chromosomes where genes lie (though Mendel had no idea what lay on chromosomes) assort randomly into sperm and eggs. Because the pattern of inheritance of ALS in the Mattingly family is straightforward, the pedigree that shows how the mutation is handed down from one generation to the next can be used to track down the defective gene responsible for the disease.

At a 1994 reunion of the Mattingly clan on Solomon's Island in the Chesapeake Bay, Andrew Mattingly Jackson saw to it that, between eating burgers, downing drinks, and heaving horseshoes, all family members had some of their blood drawn. He had the 107 blood samples delivered to Chance, by then at the University of Washington in Seattle, who extracted DNA from each sample and analyzed it to locate the defective gene.

How did Chance do this? While the process is laborious, the principle is straightforward. It's sort of like looking for a missing person. Imagine that Jimmy Bradford, a ten-year-old boy living in Marshfield, Wisconsin, is reported kidnapped by a gang of thugs who sped off in a late-model blue Chevy SUV, their current whereabouts unknown. The first thing the police do is confirm that Jimmy is not just hiding out at one of his favorite haunts: the home of his best friend, Curtis; the fishing hole he often frequents with his eccentric uncle Blake; or the game arcade in the nearby shopping mall that he visits almost daily. Finding no sign of him in any of those places, the police promptly initiate their search. They establish a five-state "Amber alert" with television and radio appeals to call a phone hotline. The next day they receive an anonymous tip: the car and boy have been spotted "about a hundred miles from Chicago," but the caller abruptly hangs up. The agent in charge interprets "about a hundred miles" to be somewhere between 50 and 150, and draws two circles on a map (see figure).

This is still a lot of terrain to cover, but it's much less ground to search than would have been the case if the kidnappers had driven for twenty-four hours in who knows what direction. The officers are pleased with their progress in locating Jimmy.

The police issue pleas to residents throughout this doughnut-shaped area to be on the lookout for the car or its occupants. Another caller to the hotline reports that she spotted a boy that fits Jimmy's description riding with two men in a blue SUV in Peoria, Illinois "not twenty minutes ago." The police quickly draw a new ring with a twenty-mile radius around Peoria and cordon off all the roads leading out of the city. With no means of escape, the thugs have to hunker down. Their SUV is spotted on Miller Lane, a quiet street in Peoria with twenty-five houses on each side. Officers go door to door, and at 106 Miller Lane they find Jimmy safe and sound and arrest his captors.

Phillip Chance and his colleagues hunted for the juvenile-onset ALS gene in a similar way. They scanned the chromosomes of the 107 family

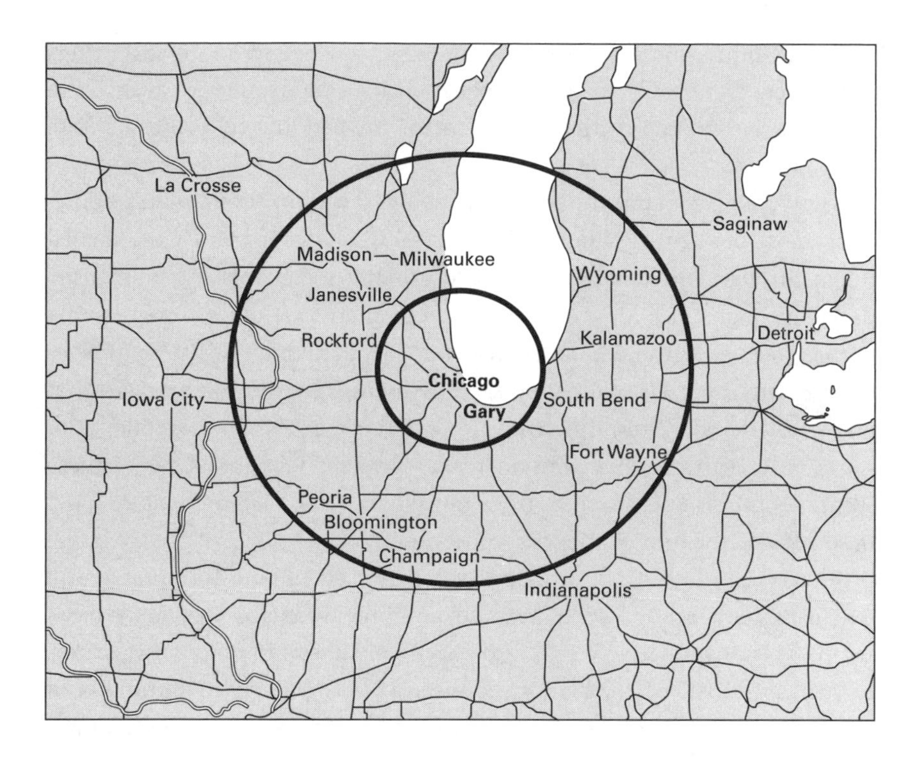

members who donated blood, looking for clues that would lead them to the defective gene. The clues they were hoping to find are particular sequences of DNA base-pairs that vary in the human population and that tend to be found in people with the disease, but not in family members free of the disease. None of these DNA sequence variants causes the disease, but some are located close enough to the disease-causing gene to provide a clue that it is nearby, much like the blue SUV on Miller Lane indicated to the police that Jimmy was nearby.

What are these DNA sequence variants and how do they point to the defective gene? They are simply small sequence differences between individuals, a reflection of that 0.1 percent difference between any two individuals' personal DNA codes. These DNA sequence variants serve as markers along the chromosomes. They are basically of two types. The first type is a single-base-pair difference, where one family member has, say, an A at a given location on one of the strands of a chromosome and another family member has, say, a G there. The second type occurs when there are slight differences in the number of base-pairs at a given location in the genome, where some family members have a few more and others have a few less base-pairs at that place on a chromosome. Most of these sequence variants lie in the 98 percent of the genome that doesn't code for a protein.

These DNA sequence differences arose by chance in the human population, most of them by about fifty thousand years ago, and were passed down through the generations to those living today (discussed in more detail in chapter 18). Most of the sequence variants are quite benign—only a small fraction of them have a consequence for anyone's health or appearance or anything else. Yet they serve a useful function as markers that allow geneticists to find their way along chromosomes. They are like signposts on a highway that announce the traveler's location. One of these signposts might tell scientists traveling down DNA: "There are 1,237,482 base-pairs to the end of this chromosome."

Identifying these DNA sequence variants used to be a formidable and expensive task, but no longer, because of impressive technological developments. It is now relatively easy and reasonably cheap to detect thousands, in fact, hundreds of thousands, of these DNA signposts in hundreds or even thousands of individuals. Many millions of these DNA sequence variants among individuals have been cataloged in the three-billion-base-pair human genome, so, like the billboards that infest many of our highways, there is a wealth of signposts along each chromosome.

How do these DNA signposts provide clues to the location of genes? One of these sequence variants, of course, is the one that causes ALS. The Mattingly family members with the disease have in their personal DNA code one particular sequence of base-pairs in the gene responsible for their ALS, which makes the gene defective. Those without the disease have a slightly different sequence at those same positions in the gene, which makes it function normally.

If the ALS mutation happened to be one of the DNA signposts that Chance scored in the Mattingly family members, he would have been able to go straight to the gene. But so far no gene hunter has been that lucky, and none are likely to be that lucky in the future. Gene hunters such as Chance don't even try to go straight to the sequence difference that actually causes a disease. Trying to find that one change in three billion base-pairs without clues would be like looking for the proverbial needle in a very large haystack; the police would have an easier time looking for a ten-year-old boy in a five-state region without any clues.

Instead, Chance looked for those DNA signposts, clues that would point him to the neighborhood of the gene. His logic was as follows. The DNA a child receives from his parents is a mix of half the DNA from each of them; this DNA is one quarter the DNA that each of the four grandparents had; one eighth of the DNA that each of the eight great-grandparents had, and so on. So the further back in generational time you go there are progressively fewer of the DNA sequence variants of each ancestor in someone alive today.

Chance knew that the region of the chromosome that contains the gene responsible for Jackson's disease must be identical in *everyone* in the family with the disease, all through the generations, all the way back to Thomas and Elizabeth Mattingly, the founders of the clan. He knew this because everyone in the pedigree who has this rare mutation and therefore the disease undoubtedly got it from his or her parent, all the way back to Thomas and Elizabeth.

Here's the key point: Some of the DNA surrounding the mutant gene is inherited along with that gene by all ALS victims through each generation. This is because the relatively few generations that have ensued since the two Mattinglys came to the United States provided relatively few opportunities for the reshuffling hatchet to separate the nearby DNA from the disease gene and cause it to be rejoined to DNA from a relative without the disease. So all the DNA sequence variants among individuals in this nearby DNA are passed through the generations *along with* the sequence difference—the mutation—that actually causes the disease. These particular DNA sequence variants are linked to that mutation, just as the specific form of the *TAS* gene that is inherited can provide information about the linked *MET* gene. Sequence variants that are linked to the disease-causing mutation thus provide clues that the gene is nearby. They travel with the mutation, just like the blue SUV accompanied Jimmy and provided a clue that he and his kidnappers were nearby.

Find those DNA sequence variants, and you'll find yourself in the neighborhood of the gene. The closer one of these DNA signposts is to the disease gene, the more often it will accompany the mutation that causes the disease—the less likely it is that the reshuffling hatchet separated it from the disease-causing mutation—and the better it will be at signaling that the disease gene is nearby.

But, just like the police first looked for Jimmy in his favorite haunts, before Chance and his colleagues began the substantial task of scanning all the chromosomes for DNA signposts associated with the disease, they first looked for them in a few places they had reason to suspect the gene might be found. They looked in the regions of chromosomes 21 and 2 that contain the *ALS1* and *ALS2* genes responsible for other forms of ALS. They looked in the regions of chromosomes 5 and 7 that contain genes responsible for a similar disease: spinal muscular atrophy. Not finding signposts pointing to the Mattingly gene in any of those places, they broadened their search and began to look for it in earnest throughout the genome.

Because they had no way of knowing on which chromosome the gene might reside, they initially cast a wide net, looking for clues throughout the genome. Since the location along the chromosomes of all these DNA signposts is known, the researchers were able to judiciously choose for their initial search a set of 150 DNA signposts that are scattered more or less evenly along each chromosome. As they got closer they could choose additional signposts in increasingly smaller regions of the genome as they tightened their noose around the culprit gene. None of these DNA signposts were likely to have anything to do with the ALS disease, but Chance knew that a few of them were bound to be situated close enough to the gene with the mutation that causes ALS to be passed along with it to the next generation.

They scored all 107 individuals for the DNA base-pair present at each of these 150 different sites, spaced at intervals of approximately ten million base-pairs across the chromosomes, looking for those positions where a particular base-pair tended to occur in people with the disease but was less frequent in people who showed no signs of the disease. Two such DNA signposts went by the names D9S158 and D9S915. Although these names may seem uninformative, to Chance and his colleagues they were equivalent to the information in that first phone call to the police about Jimmy, because they pointed them to chromosome 9 as the hideout of the disease-causing gene.

They then went over chromosome 9 with the finer-toothed comb of additional DNA signposts, looking for more clues of where the gene lies. Each time they found a DNA signpost that appeared mostly in people with the disease, it pointed them to a smaller region of the chromosome where the gene lurked. By 1998 they had limited the region where it could be to about five million base-pairs. They had tracked their prey to Peoria.

In case you're curious as to just how this gene-hunting works, we'll show you a little bit of Chance's data from his analysis of the Mattingly clan's pedigree, and work through the same logic that he and his colleagues used. If you find this whole business of walking along a chromosome too arcane, you can safely hop over this section and learn how the story ends.

The black squares and circles in the figure represent affected males and females, and the bars below two of these symbols represent a small piece of chromosome 9 surrounding the disease gene. The identities of the DNA base-pairs at different sites along the chromosome (we indicate eight of them in the figure) were determined by the researchers. For simplicity we call these sites "marker 1" through "marker 8," but Chance and coworkers knew them as D9S1831, D9S67, and so forth. The DNA base-pairs present at each site were compared to the base-pairs present at these sites in each individual's parents and four grandparents, if samples of all of them were available. Because these sites are variable from person to person, this information allowed Chance and his fellow researchers to infer the origin of each piece of the chromosome. That is, they could deduce which chromosome was maternal and which was paternal in origin; for the paternal

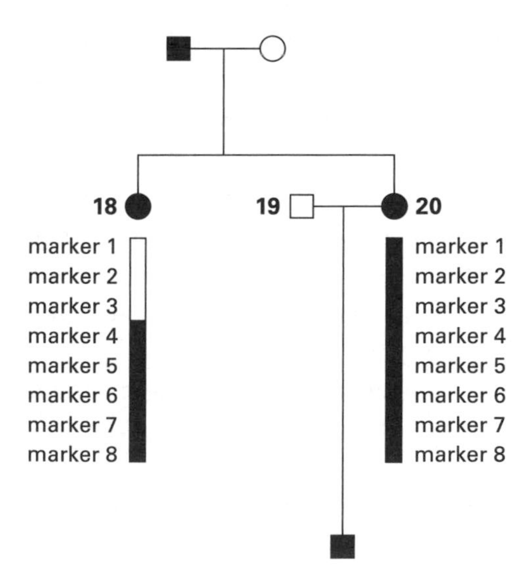

chromosome, which parts had come from the paternal grandmother and which from the paternal grandfather, and similarly for the maternal chromosome.

Two daughters, who are numbers 18 and 20 in this pedigree, are the children of a man affected with ALS. Because this father had the disease and Chance had been homing in on the relevant place in the genome where the gene must lie, the researcher knew that the man's chromosome 9 in this region harbors the mutant version of the sought-after gene. The father gave his younger daughter, number 20, this whole chunk of chromosome 9 (indicated as a solid black bar), including the ALS gene, and she contracted the disease, as indicated by the filled-in circle.

Sister 18 also inherited the disease gene from her father and she, too, suffered from ALS. But instead of inheriting this entire region of chromosome 9 intact, this daughter received a reshuffled version. In the sperm from her father that fertilized the egg that became her, the DNA on this chromosome reshuffled between markers 3 and 4. Everything south of marker 3 came from the paternal chromosome carrying the disease region; everything north of marker 4 came from the paternal chromosome with the good version of the gene.

From these results Chance could rule out the region, indicated in white for sister 18, encompassing markers 1, 2, and 3, as a possible location for the ALS gene. The ALS gene must reside south of marker 3. Continuing this process and searching for places where other reshuffling events had occurred, he could progressively delimit the piece of chromosome 9 where the gene could reside.

Chance and his colleagues continued this process over the next few years, looking for clues in increasingly smaller regions of the chromosome, until in 2002 they had it cornered between DNA sequence markers D9S149 and D9S1198, within a 500,000-base-pair region of chromosome 9. They were now on the very street that harbors the gene. And it was a small street with relatively few houses: this region of chromosome 9 contains only nineteen genes. They simply went from gene to gene, looking for a mutation—a DNA sequence difference—in one of them that is present only in people with the disease.

After two years of searching they found the mutation and captured the gene at last in 2004. The entire process took Chance and his colleagues about ten years. They could now tell Andrew Mattingly Jackson and the rest of his extended family that the mutation one of his great-great-great-great-great-grandparents brought to these shores so long ago inactivates a protein that likely works to untangle molecules of DNA or the related molecule RNA.

You may be surprised to learn that more than two thousand genes responsible for inherited human conditions have been identified in this way, and more than sixty thousand different mutations have been catalogued in them. Why do we need to know which genes cause what disease? Does it matter? Will it help us prevent or treat the disease? The answer today is an inconclusive "maybe," but we expect that it will soon be an unequivocal "yes!" for many genes.

In the case of Andrew Mattingly Jackson's ALS, the bald truth is that identification of the responsible gene in his personal DNA code has so far provided little insight into the disease. The *ALS4* gene encodes a protein called senataxin, which is similar to other proteins that are known to unwind strands of RNA (ribonucleic acid) molecules. It's not clear which RNA molecules senataxin might work on, or why a defect in unwinding RNA could lead to neurological defects. Much more will need to be learned about senataxin before Phillip Chance's success in finding the gene can be translated into real help for families with ALS4. The same can be said for most of the other two thousand human disease–causing genes that have been identified. Although the road to discovery of disease genes gets easier to travel every day, the path to understanding the disease and developing treatments and, hopefully, cures remains arduous.

There are, however, some cases where knowledge of the gene has provided real insight into the disease. Identification of the gene responsible for cystic fibrosis revealed that it encodes a protein that pumps salt molecules out of cells, which explains the buildup of liquid in the lungs of CF patients. The nature of the CF protein suggested new ways to treat the disease, some of which are currently being developed.

The identification of genes involved in certain cancers enabled the development of some highly successful drugs that destroy cancer-causing proteins. Indeed, drug development increasingly is based on our ever-growing understanding of the fundamental biological processes that go awry and cause disease, an understanding that we often gain when disease genes are identified.

But still you may wonder: Why would a medical geneticist spend a good part of ten years of his life to track down a gene responsible for a minute number of ALS cases each year? The answer is that the genes implicated in rare genetic forms of a disease may reveal something about the much more common forms of the disease that are not inherited. Indeed, this has proved to be the case time and again. Genes found to lead to rare inherited cases of Alzheimer's disease, diabetes, cancer, heart disease, hypertension, and other illnesses have told us much about what goes awry in the common cases. Phillip Chance is hopeful that the *ALS4* gene will eventually provide similar insight into the more frequent forms of adult-onset ALS.

But even if fundamental insight into the disease does not prove to be forthcoming, pinpointing the gene and the specific mutations in it responsible for the disease enables doctors to determine whether one of us has the mutation. With that knowledge, genetic counselors can then inform us of our risk of developing the disease, counsel us on how to reduce that risk, and tell us the chances that we'll pass the defective gene on to our children. In the future, knowing the defective genes that cause disease may provide a cure through gene therapy: giving to cells a functional, "good" gene. The gene-finding business has gotten a lot easier for researchers because of the Human Genome Project, and that's good news for all of us.

14 Signposts for Common Disease: Focusing on Macular Degeneration

The life of Henry Anatole Grunwald revolved around words. Grunwald left Nazi-occupied Vienna in 1938 as a teenager and made his way to New York by way of France, Morocco, and Portugal. He joined *Time* magazine as a copy boy and rose to become managing editor by the age of forty-five. He ran the magazine for the next nine years, and then moved on to become editor in chief of all Time Inc. publications for another eight years. During Grunwald's managing editorship, *Time* helped set society's agenda by adding regular sections to the magazine such as "The Sexes," "Behavior," and "The Environment," and by featuring controversial articles such as the famous 1966 cover story, "Is God Dead?". Editing was so engrained in Grunwald's blood that he usually read with a pencil poised to make revisions. He would correct any mistakes of grammar or usage he spotted, even those in published articles or books. No piece was safe from his editorial eye—once at a funeral he caught himself fixing a typo in a hymnal.

In 1973 *Time* was one of many national publications that helped solve a mystery then gripping the country, uncovering hidden clues, revealing suspicious connections, and setting puzzle pieces into place. This story involved no ordinary individuals—a break-in and its subsequent cover-up went to the highest level of the American government. As the Watergate saga played out, the case against Richard Nixon strengthened each day. Should *Time* take a stand and demand his resignation? The magazine had traditionally eschewed editorials, but this was no ordinary matter. Grunwald decided that an exception was warranted.

"How strange, I thought, that three decades ago I had arrived in this country as a young refugee, in whose eyes the figure of the president of the Unites States was, if not God-like, certainly exalted," Grunwald later wrote in his autobiography. "And here I was now arguing for a president's resignation." After going through what seemed like endless drafts, the editorial appeared in the November 12, 1973, issue. Grunwald told America that Richard Nixon "has irredeemably lost his moral authority, the confidence of most of the country, and therefore his ability to govern effectively. . . . The wise and patriotic course is for Richard Nixon to resign." The editorial drew wide public attention and provoked twenty-five hundred letters, most of them critical. Even Clare Booth Luce, the widow of *Time's* founder, wrote to complain. Nine months later Nixon resigned.

Grunwald had the courage to confront the president of the United States. A few years later he would call on his courage to confront a debilitating disease that for him was particularly poignant—a disease that likely had its roots in his personal DNA code.

•

Chances are you don't worry about contracting phenylketonuria, cystic fibrosis, or Huntington's disease. Statistically there's little chance that there is a history of any of these in your family, and even though mutations in the cystic fibrosis gene are prevalent as mutations go, it's still the case that such diseases are rare. Most likely you don't worry about suddenly coming down with any of the thousand-plus other genetic diseases whose simple inheritance patterns we understand and whose underlying genes have been identified, because they are also rare, some of them affecting fewer than one in one hundred thousand newborns.

As you age, what *are* the diseases that most concern you? Probably they include heart disease, cancer, diabetes, Alzheimer's, Parkinson's, and stroke—one of which is likely eventually to kill you—and macular degeneration, an eye disorder that makes it difficult to see fine details. The condition affects the macula, the part of the retina responsible for central vision, and its prevalence increases steeply with age. Ultimately it robs many elderly of their sight. And what medical conditions are you concerned about for your children? Probably common disorders such as attention deficit disorder and learning disabilities, depression and dyslexia, autism and asthma.

Does your personal DNA code have anything to do with common diseases like these? Unquestionably it does. But unlike traits with simple (Mendelian) inheritance patterns, these common afflictions have a complex genetic basis, for several reasons.

First, the diseases themselves are heterogeneous, meaning that they take many forms. For example, there are "wet" and "dry" forms of macular degeneration. There are also many different types of cancer and heart disease. Unlike the case with the simple hereditary diseases such as PKU, cystic fibrosis, or Huntington's, which make their presence clearly known, it may be difficult to determine whether you even have one of the common ailments. For example, blood pressure and blood-sugar levels range along a continuum from too low to too high. Where on that continuum is the cut-off for saying one has diabetes or hypertension?

Second, several different genes contribute to your overall risk of these diseases, not just one. This feature of common diseases means that there will be no simple inheritance pattern that can be determined from a pedigree such that one out of two, or one out of four, children can expect to be affected when one of the parents carries a mutation.

Third, each susceptibility gene usually makes only a minor contribution to the condition, with the extent of these contributions differing from gene to gene, and from mutation to mutation within that gene. A DNA sequence variant in one gene could make you 2.5 times as likely to get diabetes, a variant in another could make you 1.6 times as likely, and a variant in a third could make you 2.1 times as likely. Hardly the all-or-nothing effect of the defective or normal *PAH* or *CFTR* or *HD* genes. When it comes to common diseases, many individuals may have the disease but not carry a particular variant of a contributing gene; many others will carry the variant of that gene but not show the symptoms of the disease.

Fourth, genes involved in the disease may interact with one another in ways that make certain combinations of variants much worse than if their effects were simply additive; other combinations may attenuate the disease risk.

Finally, genes that contribute to complex diseases interact with the environment. For example, the effect of a variant that increases your risk of cancer may be hastened if you smoke. Conversely, the effect of a variant that increases your risk of heart disease may be diminished by a healthy diet and a cholesterol-lowering drug.

At the age of sixty-nine Grunwald had retired from *Time* and had completed a stint as ambassador to his native Austria. He was still vigorous and in the midst of writing his autobiography. While vacationing with his wife in a villa outside Florence, he picked up a carafe to pour himself a glass of water but instead poured a puddle on the table, missing the glass entirely. Returning to New York to get his vision checked for what he thought would simply lead to a prescription for new glasses, he realized that he could see virtually nothing through his left eye. The diagnosis: age-related macular degeneration. The prognosis: a continuing but unpredictable decline in sight. Such progressive loss of sight is a disturbing and disheartening occurrence for anyone; for a man whose very identity depends on reading and writing, it can be devastating.

Macular degeneration, the most common cause of blindness in the United States, takes an enormous toll: nearly one in three Americans over the age of seventy-five are afflicted with it. The disease gradually destroys the central vision needed for seeing objects in fine detail. It affects the macula, a site in the center of the thin layer of tissue at the back of the eye called the retina, which is responsible for converting light into electrical signals that are relayed to the brain, where they are interpreted as images. The "dry" form of the disease—accounting for 85 percent of cases—occurs as cells in the macula break down, leading to blurred vision. But this form can progress to the more serious "wet" form, caused by abnormal blood vessels that leak, resulting in severe damage to the macula. The wet form, the kind Henry Grunwald suffered from, can progress rapidly and lead to significant vision loss.

To track the progress of macular degeneration, ophthalmologists have patients peer at a card with an Amsler Grid (see figure), a grid of lines like those on a sheet of graph paper but with a dot in the center. Vision loss

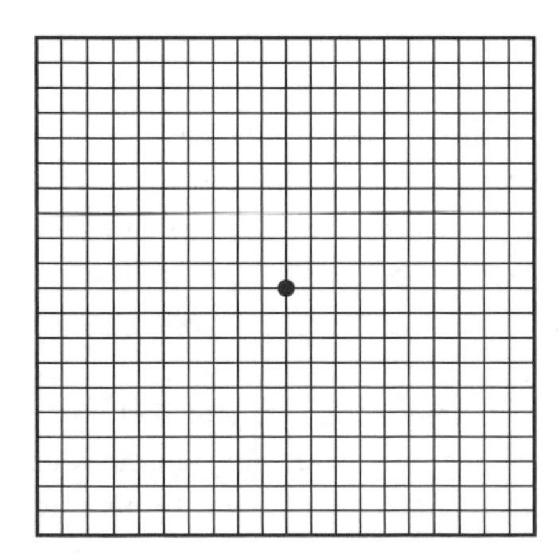

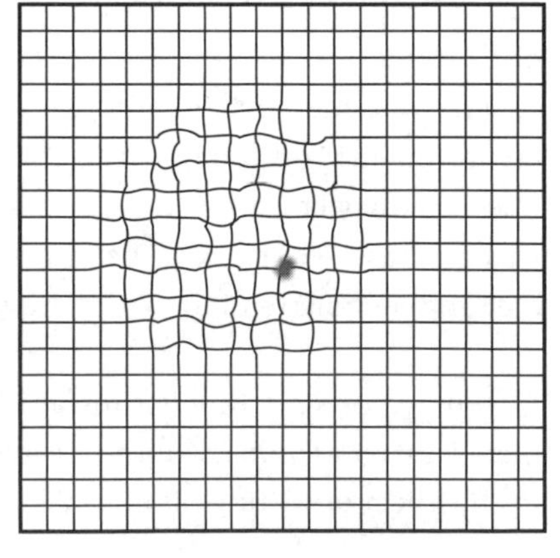

is indicated if the patient sees lines that look curvy or cannot see the dot. Grunwald described his experience with the disease in his book *Twilight: Losing Sight, Gaining Insight*: "For several months, the lines did not move, but then, just as I was becoming confident that my right eye was safe, the lines bent as if seen through heat waves. I rushed to my doctors, who spotted some bleeding in my right eye."

How do geneticists find genes that have only small effects on disease risk, but that are still critically important for our health? Family pedigrees generally won't help, because the likelihood that children will display the same disease patterns as their parents is low. The effects of the gene variants is simply too small for that approach to work.

Instead of analyzing pedigrees, geneticists compare the personal DNA codes of two groups of unrelated individuals: those who clearly have the disease, called cases, a group in which DNA sequence variants that contribute to the disease are likely to be overrepresented, and individuals who clearly do not have the disease, called controls, a group in which these sequence variants are less likely to be found. Because these kinds of DNA sequence variants do not have an all-or-nothing effect, statistical tests are needed to reveal an association between particular DNA markers and a disease.

In the figure, the white balls and black balls represent two different DNA bases at one position in one DNA strand of a chromosome—say, white balls are A bases and black balls are G bases. If we examine thirty cases and controls for some disease and ask what base they have at this position, imagine we see the distribution shown here:

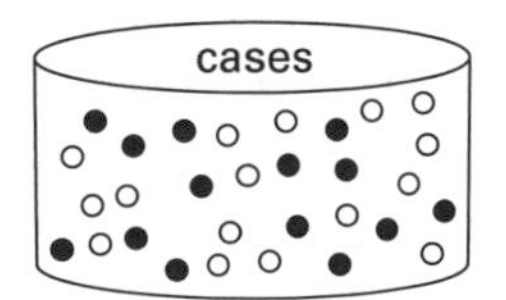

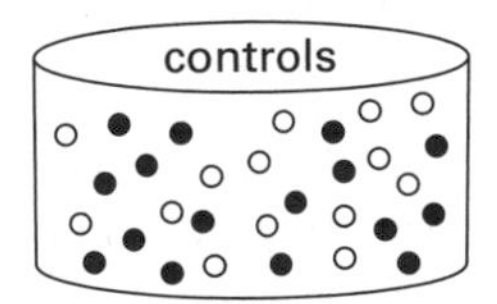

Because there are about the same number of black and white balls in each group, we conclude that the particular base variant that any individual has at this position is unrelated to whether or not he or she has the disease.

Now let's move along the genome to another position, and carry out the same analysis. In this position a white ball means a G and a black ball a T:

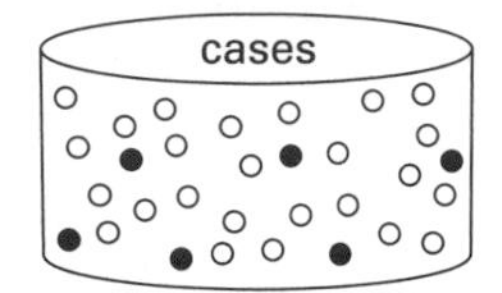

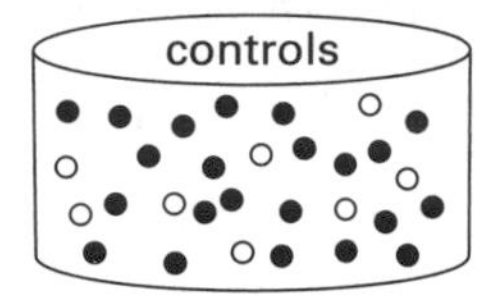

Here we see a strikingly different result: in the cases there are twenty-four white balls and six black ones; in the controls, there are eight white balls and twenty-two black ones. What these two distributions indicate is that someone with the disease is more likely than chance to have a G at this position; someone without the disease is more likely to have a T there. But note that the association is not absolute: some individuals have the disease even though they have a T (cases showing black balls), and some have a G and don't get the disease (controls showing white balls).

Robert Klein and Jürg Ott at Rockefeller University and their colleagues at Yale University School of Public Health and the National Eye Institute identified ninety-six people with cases of age-related macular degeneration and fifty control individuals with normal vision who were older than the cases (to increase the likelihood that they were truly free of the disease). Importantly, only individuals who identified themselves as "white, not of Hispanic origin" were chosen for the study, to reduce the chance that DNA differences between the groups due to different ancestries would be mistakenly linked to the risk of macular degeneration.

Then came the hard part. For both the cases and the controls, the researchers looked at 103,611 different positions in their personal DNA codes where they knew that most people have one of two or sometimes three bases on one of the strands of a chromosome—say, a G in some people and a C or sometimes a T in others. In more than fifteen million tests, they determined which base is at each of the 103,611 positions in each person, an experiment that only recently became possible with the determination of the complete sequence of the human genome, along with remarkable advances in gene-testing technology.

They posed the simple question: Is the identity of the base at any position in the genome associated with the disease state? Do people with macular degeneration have a significantly higher likelihood of having a certain base in a certain position? Looking through that haystack of 103,611 positions in the genome, the researchers found the needle they were looking for: one position, which goes by the name rs380390, was "significantly" associated with the disease state, meaning that the likelihood of this association occurring by chance is less than one in two hundred.

When Klein and his coworkers looked closely at the consequence for the health of their subjects of this single variation in the six billion base-pairs that make up the personal DNA code, they found something remarkable: people with a C at that position in one of the strands of both of their chromosomes were 7.5 times more likely to have age-related macular degeneration than someone with a G at that position in the same strand of both chromosomes. Those who had inherited only one chromosome with a C at this position had 4.5 times the risk of getting macular degeneration. Based on how often the C occurs at that position in the population, the researchers estimated that about half of the risk of developing macular degeneration comes from having inherited DNA with that base at that single position! That C does not *cause* the disease but it is a marker—a signpost—that indicates that a disease-related gene is nearby.

Grunwald went under the laser knife and emerged from the operation with stabilized vision, but five months later his sight again deteriorated. More laser treatments followed, but his macular degeneration relentlessly reduced his vision until he lived "in a half-veiled world" in which everything is seen "as if through a scrim." Grunwald eventually adapted to that world, but he chafed against its limitations. He had to resort to asking for help to cross streets, find airport gates, and press elevator buttons.

He purchased all of the various lenses and cameras and magnifiers that promised to enable some semblance of reading. In the James Bond movies, "Q" is the inventor who provides 007 with the latest fancy devices. Grunwald recalled a physician at the Lighthouse, an organization whose mission is to provide services for the blind and partially sighted, who was "Q" for the vision-impaired. The physician always had "something new to show you," proudly demonstrating to Grunwald his latest monoculars, binoculars, and extra-thick glasses. But even with such help, the erstwhile editor rued his inability to perceive subtle indicators of emotion in people's expressions, to appreciate works of art, and simply to distinguish his food in a poorly lit restaurant. Arriving at a Chinese restaurant, he attempted to shake hands with a statue of a monkey that he mistook for the maître d', provoking his son to ask, "Who's your friend, Dad?"

Just as with pedigree studies, the key assumption of association studies is that an individual with a DNA sequence variant that contributes to his disease will have inherited not just that variant, but also DNA sequence variation in the chromosomal neighborhood. We assume that more or less everyone who has some DNA sequence variant that disposes them to get a disease inherited it from the same common ancestor who lived in the distant past. The variant in our DNA was passed down through the generations, and is now common to a large number of people, at least 1 percent of us.

Klein and his colleagues reasoned that with a pedigree, because we're looking at only a few generations, the amount of DNA reshuffling that has occurred is limited. In the figure, the white rectangle is a chromosome your paternal grandmother inherited from her mom, and the black one

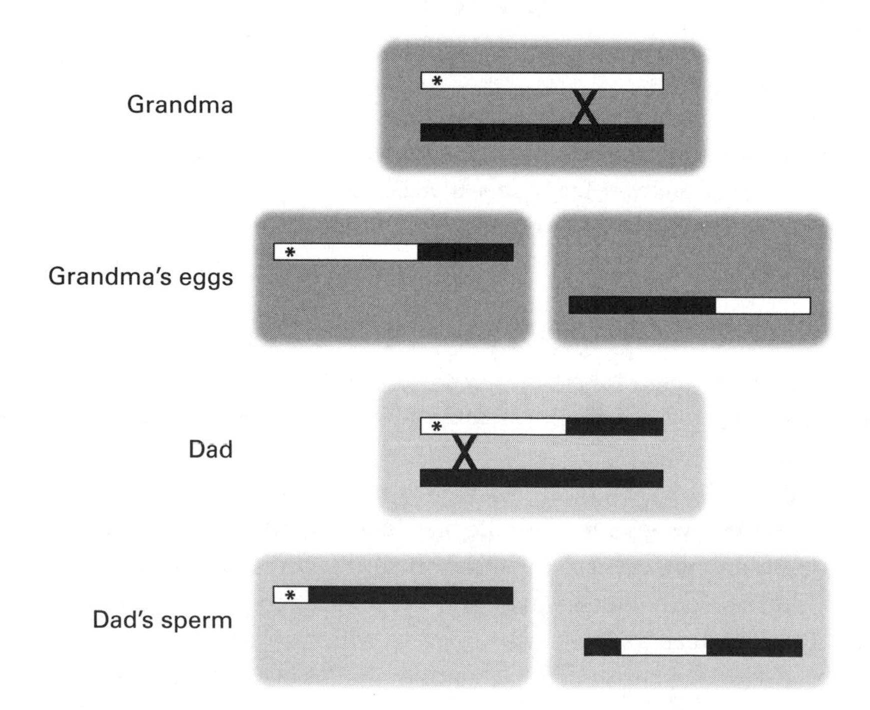

represents the equivalent chromosome she inherited from her dad. Now, let's say Grandma had a variant in one of her genes, marked by an asterisk. This variant, because it's on the white chromosome, came from her mother.

When Grandma went through the process of generating her egg cells, a reshuffling event, as described in chapter 12, occurred on this chromosome at the location marked by the X. This event resulted in egg cells with chromosomes that are a mixture of Grandma's maternal and paternal DNA, as shown.

Your dad resulted from fertilization of the egg on the left, so he inherited from Grandma the chromosome whose left half is white and right half is black. He would have inherited not only the variant gene but also about half of that chromosome's worth of flanking DNA, about fifty million base-pairs, intact from his mother. From his father he inherited the all black chromosome.

When your father produced his sperm cells, another reshuffling event occurred, again designated by an X. If you resulted from the sperm cell on the left, you inherited from him the chromosome still containing that very same variant from his mother. Now, however, the variant is in the midst of a smaller block of Grandma's chromosome, shown in white, of about 20 million base-pairs.

Children and grandchildren tend to inherit from their parents and grandparents very large blocks of DNA, amounting to millions of contiguous base-pairs. But individuals plucked from the general population are much more distantly related than that, because there were many more generations back to their common ancestor, who, as we'll learn in chapter 18, lived in Africa about fifty thousand years ago.

Each of the approximately twenty-five hundred generations that gave rise to the people in Klein's study provided an opportunity for the chromosomes to reshuffle. And reshuffle they did. Promiscuously. The reshuffling hatchet fell with each generation, and each time had the chance to separate a base-pair that varied in the population from the base-pair that caused the disease.

Each reshuffling event reduced the size of the block of base-pairs that were co-inherited. But twenty-five hundred generations is actually not all that many—taking us about fifty thousand years into the past—so the blocks of DNA that are inherited together are long enough, usually thousands of contiguous base-pairs, to include in one block both the rs380390 variant and the nearby mutation that disposes carriers to macular degeneration. This number of generations is too few to completely scramble the blocks of DNA sequence, and thus a block of co-inherited DNA—a segment in which the reshuffling hatchet never landed in those 2500 generations—remains surrounding each common DNA sequence variant. The figure shows the results of this process in this chromosome:

Although smaller and smaller segments of DNA flanking the variant (with the asterisk) are co-inherited in each generation, a block of thousands of base-pairs will remain intact, passing down to each subsequent generation the same DNA sequence variants in that region that were present in the ancestor in whom the potential macular degeneration–causing mutation arose long ago. The rs380390 DNA sequence variant, whose identity the researchers know, serves as a signpost of the disease-causing mutation because both of the affected base-pairs are in the same co-inherited block of sequence. The rs380390 DNA sequence variant serves as a surrogate for the mutation that actually contributes to the disease, indicating its presence and thereby warning of the risk of disease.

How do geneticists find such a DNA sequence variant that predicts disease risk? They use hundreds of thousands of signposts along the chromosomes that are the sites in the human genome where the base-pair sequence has been found to vary. They read what is written (A or T or G or C) in one strand on each of those signposts along all forty-six chromosomes in each of the people in the case and control groups. There are so many signposts that one is likely to lie within several thousand base-pairs of every disease-disposing mutation. An enormous set of data is produced for each individual—say, a G at base 5,268 of a certain strand of chromosome 1 in some individuals, a T at that location in others, and so on for hundreds of thousands of other locations—which is then statistically analyzed to see if any particular bases at any of the chromosomal positions appear more often than one would expect by chance in people with the disease. If one of them does, it is a signpost that signals the presence of a nearby mutation that increases the risk of disease.

Henry Grunwald lived with macular degeneration for more than twelve years, until his death of heart failure in February 2005, at the age of eighty-two. He confronted and mostly overcame the denial, the anger, and the depression that are typical of many people facing personal misfortune. But he occasionally hurled a magazine to the floor, "when, for the thousandth time, I realized that I could not read print without magnification, and I have cursed my various magnifiers as clumsy and inadequate." He turned increasingly to recorded books, to a computer that read back to him in a synthetic voice the pages he fed into a scanner, and to his wife, who read to him in a natural voice. To continue writing, he had to rely on dictation.

As his vision faded he took some comfort in mental pictures. Indeed, early in the disease he wrote, "I became a visual glutton, devouring the images around me in order somehow to hold on to them before they grew even dimmer. The faces of people I loved—my wife, my children, my grandchildren, many friends. . . . The Manhattan skyline in cold, pure autumn light. . . . Videos of favorite films. . . . Old photo albums illustrating my life. The red sunset over Vineyard Sound seen from my summer house. A Thanksgiving turkey ceremoniously displayed on a platter, a plate of pasta sprinkled with flakes of white truffles." And he came to realize that more than raging against his declining vision, he was raging against aging and his ever-nearing encounter with death. Geneticists will likely find DNA signposts for longevity, but a cure for aging will probably remain elusive.

15 The President Who Swallowed Rat Poison: Preventing the Next Heart Attack

Science progresses as much by serendipity as by deductive reasoning. Discoveries arise from a confluence of chance observations and carefully controlled measurements made by single-minded visionaries who are driven to spend their days and nights in the laboratory. Nowhere is this interplay of the planned and the providential better seen than in the story of hemorrhaging cows, a potent rat poison, a failed suicide attempt, and a president's heart attack—all part of the history of a prescription drug taken every day by more than two million Americans. New treatment strategies with this popular but potentially dangerous medicine are in the vanguard of a burgeoning field called pharmacogenomics, which promises treatments tailored to each person's unique personal DNA code.

Across the prairies of North Dakota and Alberta in the winter of 1921 and 1922, cattle were dying in large numbers from a strange disease. Autopsies revealed severe bruising and unusual swellings under the skin that were filled with blood. Minor procedures such as ear notching led to wounds that did not heal and caused the cattle to leak blood until they died. Entire herds were struck by the mysterious bleeding disease that relentlessly followed its course of thirty to fifty days and ended in death. Farmers who struggled to make a living under the best of circumstances were devastated by this disastrous turn of events.

Frank Schofield, a veterinary pathologist who had emigrated to Canada from England, was the first to decipher the basis of the mysterious outbreak. Initially he looked to bacterial infection for the cause, but he quickly ruled this out because of the lack of a fever in the affected animals, no signs of a bacterial pathogen, and his inability to transfer the disease from sick animals to healthy ones.

By 1922, Schofield had concluded that the sickness was due to sweet clover hay that had turned moldy and was being consumed by the cattle. Indeed, the harvest season of 1921–1922 had been notable for its dampness, which promoted the growth of mold. Lee Roderick, another veterinary pathologist who independently analyzed the outbreak, reported in 1929 and 1931 that the dying cattle were severely deficient in prothrombin, a blood-clotting factor. But the only treatment to arise from the veterinarians' work was transfusion of the sick cows with blood from healthy ones, and the only advice to heartbroken farmers was to find another source of hay. The disease continued to ravage the herds.

In February of 1933, Ed Carlson, a farmer from Deer Park, Wisconsin, drove the 190 miles to Madison in a raging blizzard. Sweet clover disease had killed two of his young cows in December, one of his favorite old ones in January, and two more in February, and his prized bull was "oozing blood from the nose." Carlson headed to the University of Wisconsin's Agricultural Experiment Station. Finding it closed, he entered an unlocked door he came upon and found himself in the laboratory of Karl Link. There, he dumped in front of Link a dead heifer, a milk can of blood that would not clot, and one hundred pounds of spoiled sweet clover.

Link was an agriculturist, a consultant who advised farmers on such concerns as soil quality, crop choice, and animal breeding and disease. A few months earlier, Link had been offered a faculty position at the University of Minnesota by Ross Gortner, chairman of the Department of Biochemistry. For a research topic, Gortner suggested that Link try to identify the agent responsible for sweet clover disease, and pointed him to the publications of Lee Roderick. Instead, Link had accepted a position at the University of Wisconsin in Madison, where he had decided to tackle a different but related problem. Although sweet clover *smells* sweet because of the presence of coumarin, the chemical that produces the characteristic odor of new-mown hay, in fact, this chemical makes the clover *taste* bitter. Link, assuming that the cow is something of an epicurean and prefers to eat the less bitter plants first, set out to develop a strain of sweet clover low in coumarin, to make it more appetizing to the cows.

When confronted with the catastrophe besetting Ed Carlson, Link could do no more than to tell Carlson to stop feeding his cows the spoiled hay and transfuse the sick ones, options that were not available to the poor farmer. But Link immediately changed his research goals, turning to the problem that the chairman in Minnesota had urged him to tackle: isolation from sweet clover of the "hemorrhagic agent" that was causing the cows to bleed to death. It turned into a six-year effort.

In June of 1939, after working all night in the laboratory, Link's associate Harold Campbell finally had a pure preparation of the hemorrhagic agent. Its analysis showed it to be dicumarol, a compound that results when a coumarin molecule is linked to another coumarin molecule by an enzyme present in the fungus that causes hay mold. The requirement of the fungal enzyme for this process solved the mystery of why only hay that had gone moldy caused the disease.

Dicumarol proved to be an anticoagulant, a compound that prevents blood from clotting by inhibiting a step in that process that requires vitamin K. A high level of dicumarol in the blood leads to uncontrolled bleeding caused by the thinning of the blood, which leaks out of the blood vessels. The cows were bleeding to death from the inside. Link found that dicumarol is also effective in people, and that a high level of vitamin K is an effective antidote in case of an overdose. After successful clinical trials in the early 1940s, dicumarol began to be prescribed for patients who had heart attacks, but its use never became widespread.

In late 1945, Link suffered a recurrence of an earlier tuberculosis infection and was confined to bed for eight months to recuperate. Rather than just "vegetate like a topped carrot," Link later wrote, he decided to read, of all things, "the history of rodent control from ancient to modern times." Mulling over this history, he had the stunning realization that an anticoagulant like dicumarol might make the ideal rat poison: it has no taste or odor, it is effective in small doses, and it is water-soluble and stable in cereal grains. Most critically, it does not cause immediate symptoms. This last feature of a poison is important because rats are not dumb: if one eats a bait that causes it rapidly to get sick, it makes an immediate connection and thinks: "I'm not eating any more of *that*!" Even more troublesome for a rodenticide developer, when rats see their buddies lying dead next to poisoned food, they think: "Maybe that stuff is not so good to eat." But an anticoagulant-treated rat would die several days later at a location distant from the bait, and not even the smartest rodent has the wits to think: "It must have been that dinner he ate last Wednesday."

Link appreciated that dicumarol might not be the most effective anticoagulant of its type, so he had more than a hundred different chemicals similar to dicumarol synthesized. When he returned to the laboratory the next year, he had the various derivatives tested for their anticoagulant activity on rabbits, rats, mice, and dogs. As a result of these tests, he selected compound number 42, which was much more potent than dicumarol, as the best choice for a rodenticide. Link assigned the patent rights to the Wisconsin Alumni Research Foundation, which had supported his research, and appropriated the initials of the foundation, "warf-," plus the suffix "-arin" (from coumarin), for the name of the compound: warfarin (not, as many think, from "warfare" against rats). In the decade after 1950, more than 140 million pounds of bait containing warfarin were sold under trade names such as d-Con and Rax.

We don't know why Link chose to use his convalescence to read up on rat control rather than on the origins of cancer, the building of the pyramids, the history of the Civil War, or any number of other topics. But we suspect that he already had an inkling that his long-sought and now purified hemorrhagic agent that sickens cows might somehow prove of value in the task of killing rats. Many of the best ideas in science—maybe in most disciplines—lie at the interface of apparently unconnected fields, and become apparent only when someone has the imagination to make the (seldom obvious) connection.

Today's scientists must swim through an ocean of scientific literature of a size that Link would surely have found unfathomable, and we fear that this vastness is causing many scientists to miss some crucial connections. As part of their training, all scientists are urged to read widely in the scientific literature so as to gain the knowledge necessary to recognize that the oddball fact discovered in one study is relevant to a seemingly completely different problem. But that is now almost impossible. Even in scientific journals that publish reports in just a narrow area of study, only the occasional article relates directly to a scientist's particular research focus; the rest are like the descriptions of rat poison to a group trying to cure cattle of a mysterious bleeding disease. To make sense of this prodigious scientific literature, biologists now sign up for services that email them a notice that a paper relevant to their research has been published and is ready for download. It's an efficient means to keep up, but we suspect that opportunities are being lost to make that next crucial connection that results in an important new research direction.

Ever eager to see novel applications of his research, Link surmised that warfarin, his wildly popular rat poison, might work in humans as a better anticoagulant than dicumarol. But, Link wrote, whereas "in 1941 the clinicians had literally snatched the 'cow poison' from us, . . . the transition to a substance originally promoted to exterminate rats and mice was a bit more than they could accept with real enthusiasm."

In April 1951, a twenty-two-year-old Army inductee was admitted to the hospital after a failed suicide attempt. He had become depressed by his entry into the service and tried to kill himself by ingesting rat poison: over a six-day period he had consumed d-Con containing 567 milligrams of warfarin. On admission to the hospital he complained only of back pain and nosebleeds, the latter symptom not surprising in view of a test showing that his blood was taking more than sixteen times longer to clot than did that of a person who did not partake of rat poison. The patient was treated with blood transfusions and vitamin K, and his blood tests showed that the clotting time became faster with each day of treatment. He was eventually released in good health. His doctors concluded that "accidental poisoning of an adult [by warfarin] is almost inconceivable" (although, as we'll soon see, they were more sanguine about the risks than turns out to be prudent).

This incident convinced doctors that the anticoagulant warfarin could be used to treat heart attack patients to prevent clotting, and it replaced the less potent and slower-acting dicumarol. In September 1955, when President Dwight Eisenhower suffered a heart attack, he was successfully treated with warfarin, and the publicity added substantially to the drug's luster. Warfarin came to be the long-term treatment of choice to prevent clotting after a heart attack, stroke, or surgery, as well as for people with blocked arteries, with artificial heart valves, or with the irregular heartbeat known as atrial fibrillation. And it still is.

These stories make warfarin seem like a wonder drug, but in reality its use is fraught with complications, so much so that in October 2006, Bristol-Myers Squibb, which sells the drug under the name Coumadin, added a black box warning on the packaging that cautioned of possible "major or fatal bleeding." This bleeding risk often occurs at the earliest stage of treatment, when a patient is first placed on warfarin therapy. In fact, adverse issues related to treatment with warfarin are the cause of a surprisingly large number of hospital admissions.

There are three principal problems with warfarin treatment. First, the dosage range that produces the desired beneficial effect without being toxic—the therapeutic index—is quite narrow, so that there is little margin of safety: underdose with warfarin and the patient is subject to forming blood clots; overdose with warfarin and the risk of bleeding is high. The dosages that lead to these polar outcomes are not drastically different, so warfarin is said to have a narrow therapeutic index.

Second, individuals can vary by more than a factor of twenty in the amount of warfarin they need to be usefully treated: where one person requires 0.5 mg of warfarin per day to achieve a desired anticoagulant effect, another might need twenty times more, or 10 mg, to get the same level of anticoagulation.

Third, warfarin interacts with a large number of other drugs, many of which are common and most of which increase its anticoagulation potency. Since about 30 percent of Americans over the age of sixty-five, and even more of those over seventy-five, take five or more prescription drugs per day, you can appreciate the magnitude of this problem.

The primary means for a physician to decide what dose to give a patient embarking on warfarin therapy has been trial and error; the doctor begins with a standard dose, tracks the time it takes for the blood to clot, and adjusts the dose as necessary to get it in the right range. The problem is more serious than the fact that most people don't enjoy serving as a pincushion while the correct dose is established. Individuals highly sensitive to the drug have a high risk of experiencing incidents of bleeding that can be serious, even fatal, during the first month of treatment, while the correct dose is being established.

There may be a better way, one that is coming from a new field, called pharmacogenomics, that promises individualized drug treatments matched to each person's personal DNA code. What happens when you take a drug? What does the body do to the drug? How is it distributed among the various tissues and how is it eventually disposed of? The drug first has to gain entrance into the body, most commonly by swallowing a pill or liquid, although injection, inhalation, and numerous other creative strategies have also been used. Once inside the body, the drug must cross a biological barrier, typically within the small intestine, to gain access to the bloodstream. The drug must then be transported throughout the body via the bloodstream and generally must leave the blood to reach its site of action (of course warfarin's site of action is in the bloodstream). Some locations, such as the brain, are protected and do not allow access to most drugs. Along the way, the drug gets chemically modified in the body, primarily by enzymes in the liver. This chemical modification of the drug usually makes it more likely to be excreted. Finally, the drug and its modified forms are filtered out of the blood by the kidneys and eliminated in the urine.

Some obvious factors influence relative dosage sizes. It may be easy to look at a 300-pound football player and think he'll need a higher warfarin dose than a 110-pound fashion model, or that someone with a liver disease that affects the production of drug-metabolizing enzymes might not tolerate the average dose. Older people usually require lower doses of most drugs than do younger people of the same weight, and males and females often differ in their therapeutic doses.

But acting along with these visible factors are differences in patients' genes, which play less perceptible yet perhaps more relevant roles in determining the effect of the drug. In the case of warfarin, one gene that influences a person's sensitivity to the drug encodes a member of a family of liver proteins, called cytochrome P450s, that metabolize drugs and begin the process of their elimination. Some people carry a variant of this particular P450 gene that makes a much less active enzyme. As a consequence, their bodies are not as efficient at eliminating warfarin, and their risk of bleeding with higher doses is increased. Unfortunately, this variant gene explains at most no more than 10 percent of the differences in warfarin sensitivity between individuals.

In 2005, researchers at the University of Washington in Seattle and Washington University in St. Louis found another gene, this one accounting for much more—about 25 percent—of the variation in warfarin dose sensitivity between individuals. This gene encodes a protein with another mystifying name, vitamin K epoxide reductase (abbreviated to VKORC1 by geneticists), which is actually the target of warfarin. This protein is involved in recycling the vitamin K that is used in the clotting process so that it can be used again. Warfarin reduces the amount of vitamin K in the bloodstream by inhibiting this enzyme, thereby slowing down the action of several clotting factors that depend on vitamin K.

The Seattle and St. Louis team found that there are several different versions of the *VKORC1* gene in the population. The DNA sequence differences between these variants do not affect the amino acids that make up the VKORC1 protein but instead affect how much of the protein is made. Because each of us has two copies of the *VKORC1* gene, we can carry high/high, low/low, or high/low versions of the gene in our personal DNA code. The research team analyzed individuals who had all been stabilized at their own maintenance dose of warfarin to see how the two forms of the *VKORC1* gene correlate with the dosages. The team found that those with the high/high gene combination were being treated with an average of 6.2 mg/day of warfarin, those with the low/low combination were on 2.7 mg/day, and those with one of each type of gene were on an intermediate level of the drug: 4.9 mg/day. Contrast these results with a popular regimen for beginning patients on warfarin therapy that starts everyone out on 10 mg, and you can see the value of pharmacogenomics: the knowledge of an individual's *VKORC1* genes along with environmental factors such as age and body weight now allow physicians to account for more than half of the variability in appropriate warfarin dose between individuals. As a result, patients can begin their therapy with a dose of warfarin likely to be closer to their personal therapeutic range, thereby reducing their risk of overdosing.

Pharmacogenomics is in its infancy, and as this field matures, more drug treatments will be tailored to our genetic makeup. Moreover, pharmacogenomics will have a tremendous effect on the drug development process itself by enabling "personalized medicine." This is a prospect that brings up all kinds of new social and ethical issues, especially with regard to cost and privacy, that raise the stakes of the research findings.

Cases of genetic variation among individuals that affect their response to other drugs are increasingly being uncovered. The breast cancer drug Herceptin is effective in only about one quarter of patients, women whose tumors produce a protein called HER2, which promotes uncontrolled growth of the tumor. Herceptin binds to HER2 and prevents it from carrying out its function, slowing tumor growth. But other types of breast cancer are due to tumors that grow uncontrollably for other reasons; the cells in those tumors don't express the HER2 protein, so the drug is of no use and should not be prescribed for that kind of cancer.

Another area in which pharmacogenomics will have impact is drug safety and effectiveness. Some drugs build up to toxic levels in some people but not in others, prompting searches for the genes responsible for these differences. In addition to drug-metabolizing enzymes and the direct targets of drugs, other proteins whose genes are being examined include some that bring drugs into cells or pump them out, and others that bind to drugs and carry them to their site of action.

It may also be possible to identify those people in which specific combinations of drugs are likely to be toxic. Adverse drug reactions are estimated to contribute to more than one hundred thousand deaths each year in the United States, as many as are due to breast and colon cancer combined, making it the fifth leading cause of death. So the potential value of pharmacogenomics is obvious.

Some diseases that are difficult to treat, such as depression, currently may oblige patients to go through a plethora of drugs until one is found that alleviates the symptoms without toxic side effects. The sometimes torturous road to relief is likely to become easier to navigate with a map of the patient's personal DNA code.

Pharmacogenomics also promises to advance drug development. Pharmaceutical companies are beginning to test prospective medicines on individuals with a variety of known genetic backgrounds. Some potential drugs are eliminated early in the process because they have too many unwanted side effects for too many people. But by identifying a small number of people who respond favorably to a drug, these companies may be able to salvage it to the benefit of these patients, and their own profits. This grouping of patients should lead to faster adoption of drugs, because those most likely to suffer adverse effects will not end up being medicated, and compliance will be better among those who are medicated.

Pharmacogenomics also should enable preventative medicine. If you learn you are at high risk of a common but life-threatening illness because of your personal DNA code, it might be wise to begin drug treatment years before clinical symptoms become manifest.

This all sounds wonderful, but can we afford pharmacogenomics and personalized medicine? It is expensive to develop and use on a routine basis the genetic tests needed to identify gene variants. It is incredibly expensive to develop new drugs, and if they are of use to only a small segment of the population the drug companies may not be able to recover their investment. The United States is already spending more than seven thousand dollars per person every year on medical care, about 10 percent of which is for drugs. If new drugs are developed for smaller and smaller slivers of the population, their cost will need to rise to provide profit for their makers.

There is little problem with the cost of drugs when the overall benefit is large, especially when younger and healthier groups of people are the intended users. But what about the drug that works only for a small subset of those with an advanced-stage cancer and that provides only a few additional months of life? Is this a luxury our society can afford? We need to establish our medical priorities and figure out how to cover the costs of those priorities.

There are also issues of privacy. Pharmacogenomics relies on revealing increasing amounts of our personal DNA code to physicians, hospitals, insurers, and the DNA testers. Many people are concerned about potential misuse of this information, although knowledge of a person's ability to metabolize a drug may provoke less fear of discrimination than does a positive test for the mutation leading to, say, Huntington's disease. As genetic tests become more common, we applaud recently passed federal regulations that prevent the results from endangering our ability to obtain insurance or treatment.

Pharmacogenomics deals with how our genetic inheritance affects our responses to specific drugs, but the same principles at work can be applied to many other environmental exposures, and may well lead to discovery of how our genes affect our response to pollutants, to toxins, to allergens in our food, and to all sorts of other chemicals to which we are exposed. Pharmacogenomics may be only the forerunner of a much bigger field: ecogenomics, which will explain how our genetic differences affect our interactions with the environment.

IV The Gene in Evolution

16 The Law of Evolution: Darwin, Wallace, and the Survival of the Fittest

Who made the second successful ascent of Mt. Everest? Who ran the second under-four-minute mile? Who was the second African American to play major league baseball? Who was the second man to set foot on the moon, and what did he say when he arrived there? If the public takes little note of those who come in second, scientists pay even less homage to the also-ran: priority of discovery is one of their few rewards. Which is why the unusual career of the second person generally credited with articulating the theory of evolution is captivating.

Saying that someone is perhaps the greatest naturalist of the Victorian age is saying a lot, because that was a time when natural history, the study of the earth and the living things that inhabit it, was all the rage. The Victorians loved to dabble in science, which in those days consisted mostly of collecting and classifying plants and bugs and shells and fossils. Many in the burgeoning middle class made these activities their hobby. The best of their collections were displayed in the glass cases of the British Museum, where visitors thronged to view them. If this trend had continued to today, many of us might retreat to the basement to do some stem cell research or genetic engineering after a hard day at our jobs, and spend our weekends sitting in lecture halls being regaled by the latest advances in science instead of walking around manicured meadows hitting a little white ball. Amateur naturalists were as prevalent then as weekend duffers are now.

Alfred Russel Wallace was one of those amateur naturalists, perhaps the greatest of the Victorian age. Born into a solidly middle-class family in 1823, he had to drop out of school at the age of thirteen when his family fell on hard times and couldn't support him. Wallace was apprenticed to his older brother, a land surveyor living in London, where he learned that trade and often worked for the rapidly expanding railways. Working outdoors among the birds and bees stimulated his budding interest in biology. A lull in the railroad-building frenzy put him out of work, but as would happen throughout his life when things went south for him, he landed on his feet, securing a position in Leicester as a teacher of surveying and cartography.

His new position provided him the freedom to indulge his curiosity about the natural world. "There was in Leicester a very good town library . . . and as I had time for several hours' reading daily, I took full advantage of it," he wrote in his memoir *My Life*. Among the books he read were *Personal Narrative of Travels in South America*, by the German explorer and naturalist Alexander von Humboldt, and *The Voyage of the Beagle*, by a naturalist named Charles Darwin, both of which whetted his appetite for travel to the tropics.

The book that had the most influence on him was *An Essay on the Principle of Population*, Thomas Malthus's doomsday dissertation that predicted massive global starvation due to a relentlessly increasing rate of growth of the population in the face of a more modest and constant rate of growth of the food supply. Malthus foresaw in the near future a fierce struggle for survival among people, a struggle he was sure only the strongest would survive.

It was in that public library in Leicester that Wallace met someone who changed his life, and the course of history. Henry Bates was almost a carbon copy of Wallace, and they immediately hit it off. Two years younger than the twenty-year-old Wallace, Bates was an aspiring naturalist and self-taught scientist who had also dropped out of school at a young age. He had already published a paper in a scientific journal describing his study of beetles, "and also had a good set of British butterflies," Wallace later remembered.

Bates's vast collection of beetles, with their enormous variety of forms and colors and markings, ignited Wallace's smoldering interest in collecting and classifying creatures. Bates estimated there were perhaps a thousand different kinds of beetles within walking distance of his home. This diversity of life astounded Wallace, and he was exhilarated by the realization that most of it was unknown. "I at once determined to begin collecting [and] obtained a collecting bottle, pins, and a store box; and in order to learn their names and classifications I obtained at wholesale price . . .

Stephens's *Manual of British Coleoptera*. . . . This new pursuit gave a fresh interest to my Wednesday and Saturday afternoon walks into the country."

Wallace's enthusiasm soon outstripped the diversity of life, vast as it was, in England. After a day spent in the insect room of the British Museum viewing "the overwhelming numbers of the beetles and butterflies," he began "to feel rather dissatisfied with a mere local collection" because "little is to be learnt by it." He surmised that if he studied one group of insects thoroughly, "principally with a view to the origin of species . . . some definite results might be arrived at." Wallace had been pondering the big questions all of us ask: Where do we come from? Why are we here? If he was to find answers to those questions, he had to venture beyond his current hunting grounds.

So he and his buddy Bates took off for the Amazon in 1848 to pursue their hobby of collecting wildlife. You may wonder how they financed such an exotic trip. In Victorian England, collecting things was so popular that there was a good market for beetles and butterflies pinned in boxes. Since Wallace and Bates knew that the tropics were teeming with life, almost all of it unknown to science, they saw a goldmine. They retained an agent in London who would broker their bug collections and caught the next boat to Brazil.

Wallace wandered throughout the Amazon basin for four years, first with Bates, then on his own, collecting beetles and birds and flowers and fauna, exploring regions never before seen by a Westerner. He seemed to love every minute of it, recalling fifty years later "the wonderful variety and exquisite beauty of the butterflies and birds . . . as ever new and beautiful and strange and even mysterious forms, are continually met with. Even now I can hardly recall them without a thrill of admiration and wonder. . . . There are . . . few places in England where during one summer more than thirty different kinds of butterflies can be collected; but here, in about two months, we obtained more than four hundred distinct species, many of extraordinary size, or of the most brilliant colors." He felt like a kid in a candy shop.

Truth be told, he didn't love *every* minute of it. In fact, the trials and tribulations he endured were extraordinary. The bugs that provided his livelihood also brought misery, especially the aggressive black flies: "My feet were so thickly covered with little blood-spots produced by their bites as to be of a dark purplish-red colour, and much swelled and inflamed." The weather was miserable: "Day after day the rain poured down; every afternoon or night was wet." He frequently suffered from what he called "the ague"—yellow fever or malaria, possibly both together, which sometimes weakened him so severely that he could not speak nor write nor walk, and laid him up for weeks on end.

In addition to the constant threat of disease were wild animals: "Jaguars I knew abounded here, deadly serpents were plentiful, and at every step I almost expected to feel a cold gliding body under my feet, or deadly fangs in my leg." Then there were the vampire bats, many of them rabid, whose painless bite usually came while their victim was sleeping: "I myself have been twice bitten, once on the toe, and the other time on the tip of my nose."

Most days in the rainforest Wallace followed a routine of waking at six in the morning and setting off into the jungle to collect by eight. He carried a gun for harvesting birds and small beasts, and a net for snaring butterflies and other bugs. He worked hard until mid-afternoon and then headed to the nearest stream for a bath. Upon returning to camp, he "changed . . . clothes, dined, set out our insects," and, like any self-respecting Victorian "in the cool of the evening took tea." Dinner often consisted of alligator meat, sometimes of grilled red ants. And he continued to contemplate the origin of species, confident he could make progress on the problem.

Meanwhile, at Down House, his estate in Kent, not far from London, the naturalist Charles Darwin was similarly pondering the origin of species. His son Francis remembered that his father led a life of rigid routine: he breakfasted at seven forty-five, did research until nine-thirty, then retreated to the study to peruse the morning mail, which was delivered to him by his butler and read to him by his wife, Emma. If the correspondence was light, Emma would read to him from a novel while he lounged on the sofa. Back at his research at ten-thirty, he worked clear until noon, breaking for an hour-long walk with his dogs before having lunch. He read the newspaper in the drawing room, and then wrote letters by the fireplace until three. Exhausted from all that activity, he rested on the sofa for an hour while Emma read him more of the novel. After a brief stroll for fresh air, he went back to his research for an hour, then he relaxed in his study or listened to Emma read another chapter or two. After a light dinner at seven-thirty with Emma and their five children, he played two games of backgammon with Emma. Reinvigorated, he repaired to the study to read some scientific articles until nine, whereupon Emma played for him on their Broadwood grand pianoforte. The novels must have been page-turners, because Emma read yet another chapter to him before he retired at ten.

Back in the jungle, Wallace, worn down by his daily struggles there, sick of being sick, decided to go home early. He made his way back to the mouth of the Amazon and caught the next boat back to England. Relieved that he had survived the tropics, he looked forward to a relaxing trip home, with time to organize and study his collections. But that's not what he got: about three weeks into the journey his ship caught fire and sank, taking with it the precious specimens he had risked his life to acquire, along with almost all the notebooks containing the "facts" he had painstakingly procured. Wallace almost lost his life in the disaster, floating in a lifeboat on the open sea for ten days before being rescued. After briefly mourning his loss, he characteristically dusted himself off and moved on: "But such regrets I knew were vain, and I tried to think as little as possible about what might have been, and to occupy myself with the state of things which actually existed." 'Stuff happens,' he seemed to be saying.

He had no sooner set foot in England than he started planning his next adventure: to the Malay Archipelago, today's Indonesia. He did not yet understand the origin of species, and he was more determined than ever to crack the problem. In April 1854 Wallace arrived in Singapore, where he spent four months collecting beetles and becoming accustomed to that side of the globe before setting off for points more primitive. In the wilds of marshy Borneo, humid and blazing hot, with torturous terrain and disease-bearing insects, he couldn't have been happier: it teemed with life, almost all of it unknown. For a while he was collecting an average of about two dozen new species of beetles every day. A kid in a candy shop indeed!

He struck out into the jungle every day to collect, continually pondering the question: Where did all these species come from? For most of his contemporaries back in England, the answer was clear and simple: God put them there. Even if they were beginning to doubt that He did it around six thousand years before—the time the Bible said He did it—they were convinced that He had done it. The immutability of species—what we see is what He put here—was the dogma of the day. How could it be otherwise? Surely such astounding diversity couldn't have arisen spontaneously! Surely such complexity of form and function calls for a Creator!

By then Wallace was convinced it didn't. He had long been skeptical of the conventional wisdom on this matter, and the commonalities that he saw everyday among the creatures he collected convinced him that it had happened in some other way. He could see the species were related, and he came to realize they were changing. They were not "immutable." It became increasingly clear to him that he was viewing these creatures not as they were millions of years ago; he was seeing what they had become as the result of millions of years of evolution.

Evolution. It was not a new concept. The German philosopher Emmanuel Kant suggested in the 1700s that the similarity of species is so obvious that it "strengthens the suspicion that they have an actual kinship due to descent from a common parent." The French philosopher Jean-Baptiste Lamarck, one of the most famous scientists in Europe, proposed in 1809 that "the simplest of living things have given rise to all the others." More muddled versions of this opinion had been offered throughout history, back to the ancient Greeks. But those proposals had little impact, in part because the world wasn't ready for them, in part because the authors didn't express them very well, in part because no one could think of a plausible mechanism to explain how evolution could have occurred.

In 1855 Wallace joined the crowd by concluding that "every species has come into existence coincident . . . with a pre-existing closely allied species." But unlike the others, he brought the scientific method to bear on the question: he marshaled the "facts" he had been gathering in the tropics during the previous seven years to support his conclusion. His well-written, unusually clear paper entitled "On the Law which has Regulated the Introduction of New Species" was published in 1855 in *Annals and Magazine of Natural History*, a journal that was regularly read by the scientific luminaries of the day, including Charles Darwin.

Wallace's paper was a shot across Darwin's bow. The master of Down House had been striving to solve the mystery of the origin of species since his return from a trip around the world twenty years earlier. He had been the official naturalist for that expedition, a position arranged for him by one of his professors at Christ's College, Cambridge. While on the trip he frequently left the relative comfort of HMS *Beagle* to venture ashore, most famously on the Galápagos Islands, to collect "facts" about species. He returned to Down House to mull over what those facts meant for finding a solution to the mystery of species.

And he did consider it *his* mystery. After all, his grandfather Erasmus, a polymath who was an esteemed fellow of the Royal Society, the world's oldest scientific society, was well known for his theory of evolution. In fact, the term "Darwinism" had been coined to describe Erasmus Darwin's observations and speculations on that issue. Charles Darwin was keen on keeping the evolution problem in the family.

He had been pondering the "facts" for twenty years. They increasingly led him to the same conclusion Wallace later came to: species are not "immutable"; they evolve from other species. Darwin wrote voluminous notes and some essays laying out his theories of evolution, but he told only his closest confidants about his ideas. He wanted to accumulate more support for his theories before going public with them.

Darwin was working on his masterpiece, a book describing the basis of the origin of species, but it was going slowly. At a snail's pace, really. When the journal containing Wallace's paper arrived in his mailbox, Darwin had published nothing on the subject. Not a single word on evolutionary theory in the more than twenty years he had been cogitating on the problem. And then one morning a magazine carrying an article with the clearest description yet of the evolution of species appears on his doorstep, out of nowhere, written half a world away by an upstart who was not part of the "in crowd" of Victorian science. Darwin must have been concerned. Was he going to get scooped? Was his twenty years of work to be for naught? "I rather hate the idea of writing for priority, yet I certainly should be vexed if anyone were to publish *my* doctrine before me."

"Vexed." The Victorians were masters of understatement.

Darwin's friend and colleague, Charles Lyell, the founder of modern geology, a member of the Royal Society, and the acknowledged head of Victorian England's scientific aristocracy, also read Wallace's paper and immediately saw its significance, both to science and to Darwin's legacy. He headed to Down House to prod Darwin to publish something on the topic, lest Darwinism yield to Wallaceism. There was still time: both Wallace and Darwin by now understood that there was a succession of species—new ones arising through a series of subtle changes to members of existing species—yet Wallace's paper fell short of answering the big question: *How* does it happen? Successfully scaling that intellectual summit would seal one's place in history. Darwin was pretty sure he knew the answer, but he was cautious, not yet confident enough in his theory to go public with it.

Back in Borneo, Wallace was disappointed when he received no response to his paper. He knew it was good; he suspected it was groundbreaking. "What do I have to do to get those guys' attention?" he probably asked himself. Little did he know that he *had* their attention. He didn't realize that the starting gun had sounded in the race for a place in history, even though he was the one who had pulled the trigger.

It was a race between a tortoise and a hare. Lyell had convinced Darwin he should rush to print a shorter version of his book. Darwin redoubled his effort to do that, but a tortoise can only go so fast. Meanwhile, the hare was continuing to hop around the islands of the Malay Archipelago, pinning what would become one of the largest collections of insect species ever amassed by an individual, all the while contemplating the mechanism of evolution. Fortunately for Wallace, he came down with "the ague"—malaria.

This turn of events was fortunate, because it freed him from distractions: laid up with fits of fever, unable to venture outside his hut, all he could do was

> think over any subjects then particularly interesting to me. . . . During one of these fits, while again considering the problem of the origin of species, something led me to think of Malthus's *Essay on Population*, and the "positive checks"—war, disease, famine, accident, etc.—. . . keeping all . . . populations nearly stationary. It then occurred to me that these checks must also act upon animals, and keep down their numbers. . . . While vaguely thinking how this would affect any species, there suddenly flashed upon me the idea of *the survival of the fittest*—that the individuals removed by these checks must be, on the whole, inferior to those that survived. Then, considering the variations continually occurring in every fresh generation of animals or plants, and the changes of climate, of food, of enemies always in progress, the whole method of specific modification became clear to me, and in the two hours of my fit I had thought the main points of the theory.

In a flash he had found the answer to the question he had been asking since he left Leicester ten years earlier, one of the biggest questions of all: Where did species come from?

"Survival of the fittest." Now that was a new concept! Also known as "natural selection," it is a simple statement of the fact that in dangerous circumstances, only those individuals most adapted to their environment survive—and the world, with its limited food supply, fearsome predators, and devastating diseases is always a dangerous place. An individual born with a physical or cognitive difference that renders it stronger or smarter or faster or sharper-eyed—in short, better adapted than its peers to the conditions it is confronting—will be more likely to live and mate and pass on its advantage to the next generation (through, as we now know, its genes). It's such a simple concept that it's surprising it took so long to conceive of. When Darwin's buddies back in England heard it for the first time, they must have slapped themselves on the forehead, chagrined that they had missed so obvious an idea.

As soon as he could get out of bed, Wallace put his theory to paper, producing a manuscript of 4,188 words that clearly laid out the manner by which new species arise: evolution through natural selection of the fittest individuals. By then Darwin's developing book on the same subject was over 250,000 words, and he was nowhere near the end. Darwin was still not prepared to publish, but as he would soon find out, Wallace was ready to reveal their theory to the world.

It is a simple theory, with only two fundamental components: variation and selection. Variation among individuals is the grist for evolution that gets refined in the mill of selection. Such variation is always present in a population and is manifested as differences in the fitness of individuals—their ability to survive in the environment they share. Wallace and Darwin both remembered Malthus's essay, which pointed out that resources always become limiting because populations tend to expand to exploit what's available, creating a constant struggle among individuals for those resources that are soon limited. The less fit are eliminated; the fittest survive. Variation ensures that there will always be a few individuals more fit than most, ensuring that evolution is a continual process. Hence the title of Wallace's seminal paper: "On the Tendency of Varieties to Depart Indefinitely from the Original Type."

Rather than send his paper to a journal for publication, Wallace made the fateful decision to send it to Darwin. Why? Because Darwin was famous, a fellow of the Royal Society, like his father and grandfather, whereas Wallace was a nobody. His previous paper had seemed to go unnoticed. Maybe Darwin could help him get some recognition among the scientific establishment. In the letter that accompanied the manuscript, Wallace asked whether Darwin would be so kind as to show his paper to Lyell. Wallace hoped Darwin might persuade Lyell to give his paper some visibility.

After likely retrieving his heart from the pit of his stomach and recovering from the shock of seeing, right there in his own study at Down House, *his* treasured theory issuing from the pen of another person, Darwin dispatched Wallace's manuscript to Lyell and said, in effect, "What am I to do?" Undoubtedly, Darwin was "vexed." We imagine that an exchange ensued along the following lines:

Lyell: "I told you, you should have published something on your theory sooner rather than later. But, not to worry: we'll simply present your paper on this subject along with Wallace's at the next meeting of the Linnean Society of London"—an organization whose meetings were attended by everybody who was anybody in Victorian science—"and we'll present your paper first. After all, you *did* have the idea twenty years ago. That way you'll retain your precious priority as well as your integrity."

Darwin: "But I don't *have* a paper!"

Lyell: "No problem. I'm sure you can whip something up from your notes by the time of the meeting. The next one isn't for three weeks."

Darwin: "Three weeks! Are you kidding? You expect me to write up *my* theory, the one I've been formulating for over twenty years, for its first presentation to the world, in *three weeks*?!"

Lyell (calmly): "I'm sure you'll come through. Besides, you don't have much choice."

Darwin did come through, organizing and clarifying some of his previous writings, and retrieving from friends some of the letters he had written to them explaining bits of the theory. A passable version of *his* theory of evolution by natural selection was presented at the meeting of the Linnean Society in London on July 1, 1858, after which somebody read Wallace's crisply argued paper to the group—without his consent, or knowledge, and without giving him the opportunity to edit the rough draft he had sent to Darwin. A scientist who did that today—who presented another's paper without consulting him—would be shunned by the scientific community, but back then Darwin's action was seen as honorable because it gave Wallace the credit he deserved—even though reading the paper second prevented Wallace from procuring the priority he might have had. "Darwinism" would continue to be the word that described evolutionary dogma; the less catchy "Wallaceism" would not enter the lexicon. (Wallace didn't help his cause when he later published a book describing his version of the origin of species by natural selection. Its title? *Darwinism.*)

Darwin "rushed" a book on the subject into print that appeared a year and a half later—at 477 pages a mere "abstract" of the masterpiece he was still working on. It was an immediate bestseller. Would Darwin have produced this book if Wallace had succumbed to yellow fever, or to the sinking of his ship on his way home from Brazil, or to the malaria that brought him his epiphany in Borneo? Probably not, Darwin's closest confidants acknowledged. Indeed, Darwin's long masterpiece never materialized, but the condensed version, *Origin of Species,* is still in print, as is *Darwinism,* Wallace's version of the story. So the contribution that Wallace, the first person to compose a presentable theory of evolution, is best known for is lighting a fire under Darwin, stimulating him to do what was necessary to garner credit for *his* theory of evolution by natural selection. Wallace's fire resulted in the composition of one of the most influential books ever published. Not bad for a guy with a seventh-grade education and no connections.

In the 150 years since the theory of evolution by natural selection was sprung upon the world at that meeting of minds in London, thousands of scientists have followed in Wallace's and Darwin's footsteps, reinforcing the theory, embellishing it, sometimes revising it, but never refuting it. Along the way they learned that the source of the variation was changes in DNA, the mutations that affect genes and, ultimately, the proteins they encode (something Wallace and Darwin could not have known). Most of those mutations have no effect on the fitness of the individual and therefore don't contribute to the evolution of species. Some of them reduce the fitness of an individual and get eliminated from the population. Every once in a while a mutation results in a change in the function of a critical protein in a way that makes the individual better able to compete with its peers for resources, leading to the spread of the mutation through the population with every generation. Eventually, most individuals carry that mutation, and the trait it confers on them then predominates. The population has evolved. It's a slow process, but there has been a lot of time for it to proceed—the earth formed over four billion years ago—and it does happen. The astounding diversity of life on this planet that so excited Wallace is testimony to that.

Scientists studying evolution have provided a bounty of examples of the process. Take the case of a gene in vertebrates called *BMP4*. It encodes a protein involved in building bones. Changes in that protein can have profound effects on components of the jaws of mammals and the beaks of birds. In fact, changes in this gene are responsible for the radical differences Darwin observed in the beaks of different species of birds on the Galápagos Islands, birds now known as Darwin's finches.

The differences in the beak were selected for because they render the individuals who have them better able to crack a particular variety of seed, giving them an advantage over their unfortunate rivals with less well adapted beaks. The variation in beak structure, a consequence of variation in the DNA sequence of the *BMP4* gene, was sifted and winnowed by means of natural selection. A change in the *BMP4* gene that makes the beak a little thicker or a little wider or a little stronger enables the bird fortunate enough to possess that particular form of *BMP4* better able to crack open seeds with hard shells, and therefore better able to compete with its rivals for those kinds of seeds. Another bird with a variant of *BMP4* that results in more curvature of the beak finds itself slightly better than its rivals at digging into fruit for seeds.

Small changes like these in the *BMP4* gene, and many, many others in many, many other genes, went through the sieve of natural selection time and again, over many millions of years, resulting in the several very closely related species of finches that Darwin observed on the Galápagos Islands, each with its own distinctive style of beak. If you visit there today you will see the descendants of the birds Darwin studied, but their beaks are slightly different now, because they have evolved in the 179 years since Darwin set eyes on them. And the beaks of those birds are still evolving according to the principles of natural selection, which culls the persistent variation in *BMP4* and other genes in our dangerous world.

Some of the most beautiful and convincing examples of evolution began to surface about thirty years ago, when scientists gained the ability to determine the sequence of amino acids in proteins and of base-pairs in DNA, using methods developed by Fred Sanger (discussed in chapter 3). The relatedness of organisms was immediately obvious when the first short DNA sequences began to emerge from a few labs. When the sequences of entire genomes of organisms began to flow from DNA-sequencing factories a dozen years ago it became crystal clear: all organisms on the planet are built from the same set of genes. That's because, way back when, we all had the same ancestor. Its name was LUCA—for "last universal common ancestor" of all living things—and it lived around three and a half billion years ago. This organism is no longer around, and no one knows what it looked like, but we can be confident that it was a one-celled creature. And it had genes made of the exact same kind of DNA we have, and it passed them on to us and to all the other living things we share our world with. As near as anyone can tell, life arose just once on earth, and we're all descended from that event. LUCA and the next cell it gave rise to when it split for the first time are the real Adam and Eve.

So we're just going to have to accept it: our genes, which encode our proteins, are not all that different from those of the fly that just landed on the kitchen counter. About half our genes are obviously similar to its genes, meaning we can line up the sequence of base-pairs in the human and fly versions of many genes and find lots of positions where the base-pairs are identical. In fact, they're so similar that many of our genes will work in the fly: we can generate a fly with one of its genes replaced by the human counterpart and observe nothing obviously different with that fly. The experiment has been done many times.

Flies get Parkinson's disease, and it's because of defects in some of the same genes that contribute to Grandpa's Parkinson's disease. Flies get heart disease—or a disease of an organ that passes for their heart—because of defects in genes similar to some of those that probably contributed to Isaac Asimov's heart attacks. Flies get brain tumors due to defects in genes similar to some human genes responsible for certain kinds of brain tumors. We're not all that different, genetically speaking, from flies. You don't want to know how similar your DNA is to that of rats!

We even carry some of the same genes as that yeast that produced the wine you had with dinner last night; one of them is the *WRN* gene, which encodes a protein that unwinds the DNA double helix so it can be copied. Humans have virtually the same gene. It's not exactly the same, because natural selection—acting on variation in the *WRN* gene over the billion or so years since humans and yeast evolved from their common ancestor—has changed it. But the two genes still encode obviously similar stretches of amino acids in their proteins; it is clear they evolved from a common source.

```
human     MESYYQEIGRAGRDGLQSSC
dog       MESYYQEIGRAGRDGLQSSC
mouse     MESYYQEIGRAGRDGLQSSC
chicken   MESYYQEIGRAGRDGLPASC
fish      MESYYQEIGRAGRDGLPSAC
worm      IESYYQEIGRAGRDGSPSIC
fly       IEGYYQEAGRAGRDGDVADC
rice      LESYYQESGRCGRDGLPSVC
yeast     LEGYYQETGRAGRDGNYSYC
bacteria  IESYYQETGRAGRDGLPAEA
```

The *WRN* gene is implicated in a disease called Werner's syndrome, which leads to premature aging. The figure shows the sequence of amino acids of a stretch of the WRN protein in humans, dogs, mice, all the way to bacteria. Each letter is an abbreviation of the name of one of the twenty amino acids. The sequence of this stretch of amino acids in the WRN protein is identical in humans, dogs, and mice; only six of the amino acids are different in this region of the bacterial protein.

The protein encoded by the *WRN* gene seems to be doing essentially the same thing for both humans and yeast. People with a defective *WRN* gene age unusually quickly, looking as though they're eighty years old by the time they're about forty. Most people who inherit Werner's syndrome die in their forties or fifties. Remarkably, mutations in the equivalent gene of yeast have the same effect: the yeast cells age about twice as fast as normal. Yeast aging is measured by the number of "daughter" cells they give birth to, and a yeast cell with a defective *WRN* gene dies young, giving birth to about twenty daughter cells rather than its usual litter of forty. Mutations in the *WRN* gene have the same consequence in two species that are vastly different because of the natural selection that operated on variation that arose over the last billion years. Remarkable? Not really. It's what Wallace and Darwin said we should expect.

Natural selection—"survival of the fittest"—is an easy concept to grasp. And one hundred fifty years of research by many thousands of scientists have generated evidence to support the validity of the hypothesis beyond any shadow of doubt. Why, then, is the issue still so controversial? We honestly don't know why. Perhaps it's because the notion of evolution conflicts with the view that many held in Darwin's time and that some still hold today: that God created living things just as they are today. Perhaps there is lingering doubt that humans and apes are related. Perhaps clever writings designed to sow suspicion of evolution to further some religion-driven agenda are persuasive to some. We lay some of the blame on the term "the *theory* of evolution." It is a theory because no scientific principle can ever be proved: we can only disprove alternatives, at the same time postulating the most plausible version of reality consistent with our observations (we can't be completely sure that the next time we release an apple it won't fall *up*, but on the basis of the available evidence we hypothesize that it will fall *down*). In light of the overwhelming evidence that supports the theory, evolution by natural selection is right up there with the *law* of gravity, the *law* of conservation of energy, and the *law* of supply and demand. We think it's time to ditch the word "theory" and call it the "*law* of evolution."

The second successful ascent of Mt. Everest? Ernst Schmied and Juerg Marmet, of Switzerland. The second under-four-minute mile? John Landy, of Australia. The second African American to play major league baseball? Larry Doby, with the Cleveland Indians. The second man on the moon? Buzz Aldrin. His first words when he arrived there? "I'd like to take this opportunity to ask every person listening in, whoever and wherever they may be, to pause for a moment and contemplate the events of the past few hours, and to give thanks in his or her own way." We expect you are now able to correctly answer the question: Who was the second person to pen a coherent theory of evolution? Charles Darwin.

17 Around the World in Fifty Thousand Years: The Genetics of Race

The Food and Drug Administration (FDA) approved BiDil, a drug for the treatment of heart failure in self-identified black patients. . . . "Today's approval of a drug to treat severe heart failure in [a] self-identified black population is a striking example of how a treatment can benefit some patients even if it does not help all patients," said Dr. Robert Temple, FDA Associate Director of Medical Policy. "The information presented to the FDA clearly showed that blacks suffering from heart failure will now have an additional safe and effective option for treating their condition. In the future, we hope to discover characteristics that identify people of any race who might be helped by BiDil."
—*FDA News*, June 23, 2005

Top-seeded Jimmy Connors stepped onto Centre Court at Wimbledon for the 1975 men's final having declared that it would be "just another day at the office." Ranked number one in the world, the twenty-two-year-old defending Wimbledon champion had not dropped a single set en route to the final. The brash left-hander was the overwhelming favorite against the other finalist, sixth-seeded Arthur Ashe. Connors was famed for his explosive outbursts on the court; the thirty-one-year-old Ashe calmly closed his eyes and meditated between games.

The three previous times these rivals had met, Connors had prevailed decisively, and commentators at Wimbledon that day hoped only that Ashe would not be embarrassed on the court. To their surprise, Ashe began the match in dazzling fashion. Instead of trying to out-hit the hard-slugging Connors, Ashe brilliantly executed a game plan of slices, chip returns, lobs, and other change-of-pace shots, to dominate the first set 6–1. When the second set began going Ashe's way, a fan yelled out, "Come on, Connors!" and Connors shouted out, "I'm trying, for Chrissake." The crowd's laughter distracted Ashe only momentarily, as he dominated again 6–1.

Connors, never a quitter, fought back to win the third set 7–5, keeping alive his hopes for a stirring comeback. He started out the fourth set strongly to gain a 3–0 advantage and was but a point away from 4–1, but Ashe, resolutely sticking to his game plan even when on the defensive, rallied to win the set and match and become the first and still the only African American to win the Men's Championship of the All England Club. It was the only time he would ever defeat Connors.

Four years after his triumph at Wimbledon, while participating in a tennis clinic, Ashe suffered a heart attack that necessitated quadruple bypass surgery four months later. It forced his retirement from tennis soon after, and his continuing heart problems led to more surgery in 1983.

Ashe never forgot his childhood in segregated Richmond, Virginia, where he had been excluded from whites-only tennis tournaments. Or the Davis Cup match in 1965 between the United States and Mexico that had to be moved from the private Dallas Country Club to a public facility because club members objected to his presence on their courts.

Ashe once said that if he was remembered only as a tennis player he would have been a failure. But he is remembered as much more. He took a highly public stand against apartheid in South Africa, and his visit there in 1973 included the first match in a stadium with integrated seating and an integrated locker room, a sporting event that helped catalyze change in that troubled country. Following his playing days, Ashe wrote *A Hard Road to Glory*, a three-volume history of the American black athlete. Shortly before he died in 1993 at the age of forty-nine, he was arrested outside the White House for protesting U.S. immigration policy toward Haiti. That the main stadium for the U.S. Open in Flushing, New York, is named for him is not just testament to his tennis; his friend and playing partner Bob Davis said, "Arthur's name is on that stadium because of the exemplary life he led. He was a tennis champion who transcended the sport."

Few issues are as contentious in American society as race. As stellar a citizen of court and country as Ashe was, he was at one time called an Uncle Tom for appearing to legitimize the South African government. At other times he was criticized for not doing enough to further the careers of young black tennis players. Given the history of race in America, the relationship between race and genetics is a landmine for researchers who attempt to study the subject. A host of issues—the very definition of race, the dispute over whether race is a valid categorization of people, the question of which traits might have a race-specific genetic basis, the utility of using racial identity to assist in finding disease genes, and the value of targeting drugs to certain racial groups—all of these topics provoke intense feelings and heated debate. Because humans seem to have a need to define and differentiate themselves, and because many Americans believe race is so evident a category since it seems to be plainly visible in front of their eyes, the use of race as a classifier of people pervades much of our collective daily existence.

In tackling the issue of genetics and race, we are painfully aware that the widespread racial discrimination in America's history was often aided by ostensibly objective geneticists claiming to draw on the latest scientific orthodoxy. Efforts from the seventeenth century onward to classify humans into major groupings perpetuated the notion that the classifiers—invariably white men—belonged to a nobler group than did members of other races. This kind of eugenic thinking culminated in the United States with Jim Crow laws such as the "one-drop" rule, as formulated in the Racial Integrity Act, passed by the Virginia legislature in 1924: "It shall hereafter be unlawful for any white person in this State to marry any save a white person, or a person with *no other* admixture of blood than white and American Indian. For the purpose of this act, the term 'white person' shall apply only to the person who has *no trace whatsoever* of any blood other than Caucasian" (italics added). This law stood until it was declared unconstitutional by the U.S. Supreme Court in 1967.

It would be most unfortunate if recent findings from the Human Genome Project and our increasing ability to characterize our personal DNA codes led to a revival of genetic determinism based on racial groupings. Particularly misplaced is the notion that if a genetic association between a disease and a racial group is found, then all members of that group, including individuals who don't carry the gene variant disposing them to the disease, share the same risk of the disease, especially when the group at risk is not even clearly defined. Even worse is the view that some studies can be construed to support the presumption of a racially determined genetic basis for traits such as athletic ability, intelligence, or criminality, without any good evidence for such a claim.

There is no question that various communities in our society today face enormous disparities in their access to health care, education and employment, and in their diets and levels of stress. Overwhelmingly these disparities boil down not to genetic differences but to economic disadvantages: health is wealth. Yet even when the statisticians account for economic inequality in access to health care and treatments, certain diseases have a much greater prevalence or significantly more severe outcomes in certain populations traditionally viewed as races. Why is this the case?

To understand race and genetics, we have to consider where we came from and how we got here. The fossil evidence suggests that anatomically modern humans, those with physical characteristics not too different from our own, emerged in Africa about 200,000 years ago—an exceedingly brief period in evolutionary time. These humans were part of the lineage of hominids, the family of great apes that comprises humans, chimpanzees, gorillas, and orangutans. The branch of that family that leads to humans split off about six million years ago from the last ancestor we shared with our closest relatives, the chimpanzees; the chimpanzee's DNA sequence is about 99 percent identical to ours. The human evolutionary tree indicates that humans did not evolve from current chimpanzees (or monkeys or gorillas or apes). Rather, both we and chimpanzees evolved from an ancestor no longer in existence who lived about six million years ago and whose various descendants would eventually give rise to two lineages, one that became us, and one that led to today's chimpanzees.

Anatomically modern humans first appeared in sub-Saharan Africa, and groups of them ventured out of Africa around 50,000 years ago, spreading throughout the Eurasian landmass—the Mediterranean coast, Europe, Russia, and central Asia—and into Australia. They got to Siberia by 30,000 years ago, and then moved across the Bering Strait and into the Americas about 15,000 years ago, along the way inventing paper in China, mathematics in the Middle East, and country music in the United States. The key point to remember here is that humans spent about 150,000 years in Africa before they colonized the rest of the globe.

Upon their arrival in Eurasia, these early humans likely met the Neanderthals, an abundant hominid species that inhabited Europe and western Asia from about 400,000 to 30,000 years ago. Neanderthals and ancient humans last shared a common ancestor about 500,000 years ago, long before humans walked out of Africa. There is evidence that the globetrotting humans met their Neanderthal cousins in several places, but a comparison of our DNA to theirs suggests that the two groups never got to be very intimate.

Why did it take so long for early humans to venture out of Africa to enjoy the abundance of the rest of the world? It was around the time of the initial migrations out of Africa that humans acquired more sophisticated tools, ornaments, and weapons, even indulging in abstract art, all activities that were evidence of their increased intelligence. This greater brain capacity correlates with a major increase in the population during that time, which may have made them more able to strike out to find new places in the world.

The implications of this model of human evolution are profound. It means that all six-plus billion of us on earth today descended from a small number of people, probably no more than ten thousand, who lived in Africa around fifty thousand years ago. The migrants who left Africa for points distant were a small subset of all the individuals then alive in Africa, a fact that has far-reaching consequences for human genetics.

If we look at the personal DNA codes of several present-day people to see how many DNA sequence differences we find in them—that is, in how many positions in the genome one person has, say, an A on one strand, and another person has a G—we learn that the number of these variants is significantly greater among Africans than it is among people in other geographic groups. Furthermore, most of the variation seen in populations outside of Africa is also present in the people who live in Africa. For example, if we find that the base at a particular position in the genomes of some Asian people is usually a C and occasionally a T, then we typically find among the African population both the C and the T (and maybe a G as well) at that position. This is because the emigrants brought with them only a sampling of the genetic diversity in the population they left behind. In this case only people with C and T at the position in question emigrated; people with a G at that position stayed behind. Some of the variation reached locations around the globe, but all of it was left behind in the people who stayed back to hold down the fort in Africa. Of course, all humans—those living in Africa as well as those who populated other lands—continued to evolve.

The consequence of a slowly spreading human population is "Race is space," as Rick Kittles of Howard University and Kenneth Weiss of Pennsylvania State University put it. A new DNA sequence variation that arises in a single individual may spread geographically, but it will move slowly because human generations are long, about twenty years, and in our evolutionary history prior to the advent of Internet-based matchmakers we tended to mate only with our neighbors. So if a particular gene variant is found in populations around the globe, it is likely to be ancient, and was probably present fifty thousand years ago, when our ancestors hiked out of Africa.

Conversely, rare variants tend to be much more recent, meaning that they have arisen within the last fifty thousand years, and tend to be found only in individuals living in particular regions. In other words, gene variants have been accumulating in people living in Africa for about two hundred thousand years, much longer than the fifty thousand years they have had to accumulate in the population residing on the rest of the planet.

The geographic clustering of early humans did not generate discrete racial groups. Instead, the genetic variation in humans spread in gradients, with the frequency of one particular form of a gene increasing in some directions, decreasing in others. Thus the greater the geographic distance between two populations, the greater their genetic differences: the development of different races is simply due to the space seperating them that leads to two genetically distinct populations.

So in light of this history, what is race? There is no generally and consistently accepted definition. Some define a racial group according to physical features such as skin color and hair texture, which reflect a shared ancestry. But others see race as purely an invention, often of white males, to justify cultural practices. Regardless of the conflicting definitions, we think it's fair to ask whether a biological basis for the concept of race exists.

The key point to bear in mind when discussing possible biological groupings of humans is that no matter what genes you examine and no matter how you define your population groups, about 85 to 95 percent of all the genetic variation you observe in our personal DNA codes is found *within all* of the population groups; the small balance of variation is all that exists *between* groups. Imagine that we took random groups of citizens from Cameroon, China, Canada, and the Czech Republic and sequenced all six billion base-pairs of their personal DNA codes (something we'll be able to do soon). We'd find a lot of differences within each of them (the six million or so that we talked about in chapter 11). But the differences that we'd find among those from Cameroon are very largely the same ones we'd find in common among people belonging to the other three groups. Could we use these sequence data to define a gene or genes for being Cameroonian, or Chinese, or Canadian, or Czech? Of course not! No such genes exist. We are all way too similar in our genetic makeup for that to be possible. Nonetheless, if you compared the variation present in say, the Cameroonian, to the variation present in all the world's population groups, you would probably find enough specific differences to be able to place that individual quite close to Cameroon.

What about the 5 to 15 percent of the variations that have been found to be typical of one human group or another? Do some of these affect skin color or hair texture or other differences in appearance? Of course they do. All of our physical traits are ultimately determined by our genes. We humans are so anthropocentric that when we look closely at our fellow beings, we notice the tiny differences in the shape of an eye, the slope of a nose, the thickness of a lip. But climb only a hundred feet up a hill and you will have a hard time distinguishing those characteristics. From that perspective we are all nearly identical: the same size, the same shape, with the same number of arms and legs, the same locations for eyes and ears, the same everything else. Thus, classifying individuals into a few groups based on minor differences in appearance and then using those groupings to make inferences about the genetic basis of complex social behaviors is to ignore the huge amount of genetic variation everyone in the world shares.

How do the worldwide patterns of genetic variation that exist affect our ability to identify disease genes? Clearly, some diseases are more prevalent in individuals in one group than in those of another. The prevalence of Tay-Sachs disease is higher in Ashkenazi Jews than in other groups; sickle-cell anemia is most frequent in Africans; phenylketonuria is essentially absent in Africans. "We do not sample Lapps to study Tay-Sachs Disease, Norwegians for sickle cell anemia, or Nigerians for PKU," write Kittles and Weiss.

So here's the heart of the race/genetic relationship. Unless and until widespread intermarriage among all humans leads to one homogeneous population, we can more or less divide most of the worldwide pattern of local genetic variation into a few large general, and quite rough, groupings: Africans, Europeans and Middle Easterners, east Asians, and Native Americans. (This oversimplified scheme leaves out a host of smaller subpopulations.) These groupings—which you can call races if you want—contain that small percentage of rare DNA sequence variation (5 to 15 percent) that produces the diversity in the global police lineup.

More important, these rare variants contribute significantly to differences in people's risk for certain diseases. For example, African American women often develop breast cancer at a younger age than white women who get the disease, and have nearly double the rate of an aggressive form that is resistant to many treatments. Physicians who treat breast cancer patients are beginning to look to Africa to explain some of these differences, hoping to find genetic variants there that may predispose black women to this virulent form of the disease. Another example: genetic variants among Ashkenazi Jews, a small subgroup of all humans with European origins, lead to an incidence of Tay-Sachs disease one hundred times greater than is found in other populations. But these differences in our DNA don't reflect some kind of inferior genetics, any more than the much higher rate of PKU in people with lighter skin than in people with darker skin says anything about racial fitness.

What about BiDil and the targeting of pharmaceuticals to racial groups? This drug is actually a combination of two drugs, hydralazine and isosorbide dinitrate, that had been available for decades and are sold in generic form. Earlier studies of the drug combination had not produced evidence convincing enough to justify its approval, but an analysis of subgroups of patients—a suspect form of data analysis because the question being tested is stated after you have the answer—revealed a benefit of the drug for "blacks." This finding inspired a new trial called the African-American Heart Failure Trial, carried out only on self-identified African Americans. The results were stunning: BiDil, used along with conventional therapies, led to a 43 percent increase in the rate of survival of heart failure patients compared to those in the study treated only with conventional therapies.

Surely you can appreciate that BiDil does not target the products of genes that influence skin color or hair texture or facial features. Rather, some combination of differences in hypertension, salt sensitivity, and other physiological properties in this self-identified population might differ from the rest of the population such that this drug is especially effective for them. As of this writing, the specific differences in our personal DNA codes that are the basis for this difference aren't known, but they probably will be soon. At that point, regardless of your skin color or what ethnic group you associate yourself with, if you have the BiDil-sensitive variations in your DNA code, the drug will likely help you. And however dark your skin, or however closely you identify yourself with others with dark skin, if you don't have those particular DNA sequence variations in your DNA code you won't be helped by BiDil.

We don't know whether BiDil would have helped Arthur Ashe after his heart disease became apparent. It likely wouldn't have mattered anyway. Five years after his second heart surgery, Ashe was hospitalized for toxoplasmosis, a parasitic infection, and learned that he had AIDS, apparently caused by the presence of the Human Immunodeficiency Virus (HIV) in blood he received during his surgery in 1983. Ashe held a press conference in April 1992 to announce that he had the disease. A year later he was dead of AIDS-related pneumonia.

Will the knowledge of the specific DNA sequence variants each of us carry in our personal DNA codes—which affect disease susceptibility, drug efficacy, and many more things that are important to us—end the racism in America that Ashe worked hard to overcome? Will health disparities disappear because we can determine the sequence of DNA and therefore no longer need to classify individuals on the basis of appearance to take advantage of their genetic differences? Likely not. We know all too well that those societal outcomes won't be realized because of new genetic knowledge. But we can hope that genetic knowledge won't make the problems any worse. Someday, perhaps, we'll come to appreciate that even though the 0.1 percent difference in the DNA between any two of us might mean the difference between being or not being disposed to get a particular disease, the 99.9 percent similarity means that we're all close relatives: we are all descended from the same ancestors who came out of Africa not so long ago.

18 Your Personal DNA Code: Summing Up

If you've made the journey with us to this point, you've learned some little-known facts about an eclectic collection of characters, some obscure—Patricia Stallings, Mike O'Brien, and Andrew Jackson Mattingly—others world-famous—Pearl Buck, Rita Hayworth, and Katie Couric. All of them confronted the heart-breaking consequences of a small change in their DNA code or in that of a loved one. You saw that the prognosis of people who are dealing with a lethal infectious disease, like Isaac Asimov and Arthur Ashe, is affected by their genetic endowments. And you have come to know some renowned biologists—Karl Link, Frederick Banting, and Seymour Benzer—who were driven to discover the principles of life's processes, and who were subject to the same inspiration, competition, determination, trepidation, and exhilaration that drive all of us.

But the personal stories were just appetizers. Our main dish consisted of answers to the basic questions that we all have about genetics, questions we confront daily in reports of discoveries of how genes influence our lives. If you digested the meal we set before you, the following will sound familiar to you.

Each of us inherits two sets of chromosomes, one from each of our parents, which are very long strands of A, C, G, and T, the chemical units of DNA that twist around each other in a double helix. Long is truly long: one hundred million or more DNA base-pairs can be present in a single chromosome; six billion base-pairs are crammed into each one of the trillions of cells in our bodies.

The critical feature of our DNA is the order—the sequence—of its A, C, G, and T letters. That sequence of A's, C's, G's, and T's is unique to each of us (unless we're an identical twin)—it's our personal DNA code. It's what makes everyone different from the other six and a half billion people on our planet.

Our chromosomes are partitioned into about twenty thousand segments called genes, each of which provides the information to manufacture a protein. Proteins, composed of unique sequences of twenty different chemical units called amino acids, do the work in the body's cells: breaking down food, making energy, signaling the state of affairs, protecting us from invaders such as bacteria and viruses, and much more. We each look different from everyone else on the planet, and our cells carry out their affairs slightly differently than everyone else's cells, because the exact sequences of amino acids in our proteins are slightly different from everyone else's. That's because a small fraction, about 0.1 percent, of each individual's DNA code is slightly different from everyone else's.

The distinct types of cells in the human body—blood, nerve, muscle, skin, and the rest—look and behave differently because they contain distinct subsets of the twenty thousand proteins encoded in our DNA. Some specialized cells are the only ones that produce a particular protein; for example pancreatic "beta" cells are the only cells that produce insulin. All cells contain many common proteins that carry out necessary functions, such as metabolism. The kinds of proteins a cell makes is determined by which of its genes are turned "on," a decision made by a special type of protein called a transcription factor. These are important decisions indeed: beginning with a fertilized egg, each cell division brings new corps of transcription factors to orchestrate the relentless specialization of cells that results in the birth of a complete and utterly unique human being.

The precise collection of proteins in each of us is determined by the specific versions of the genes we inherit from our parents. Our proteins are more similar to theirs than to those of any other pair of parents, making us resemble our own mom and dad more than we resemble others' moms and dads. And because our siblings inherit many of the same versions of genes as we do, our resemblance to them is also usually obvious. Proteins affect more than just our looks: they are involved in learning and memory, mood and behavior, addiction and desire. Each of these traits is influenced, to varying degrees, by our personal DNA codes.

Because the process of copying DNA during each cell division is imperfect, mutations—changes in the sequence of the DNA letters—continuously occur. Some of the mistakes are replacements of one letter for another; others are insertions or deletions of letters, sometimes thousands or even millions of them. Mutations can add to the diversity of life; by changing the sequence of letters in the DNA they can change a gene, thereby altering the action or the amount or some other property of a protein. In consequence, we look or act a little bit different than we would have if no DNA copying mistake had been made. Mutations contribute to our individuality.

Although most mutations are inert in their effects, some mutations eliminate the ability of an important protein to do its job. Usually that's not a problem, because most of these mutations are recessive, meaning that with two sets of chromosomes we have a backup copy of every gene, and the backup is sufficient. These recessive mutations make mayhem only when they are in a double dose, one coming from each parent. The consequence can be a terrible disease if the missing protein is needed to carry out an essential task. Occasionally a mutation creates an altered protein that all by itself becomes a wrench in the gears of our cells. A single copy of such a dominant mutation is sufficient to lead to disease. When one of Mom's or Dad's chromosomes carries a mutation, what determines whether we get the good copy of a gene or the bad one? Only blind luck: the decision is determined by which of Dad's millions of sperm cells fertilizes which of Mom's egg cells. It's an important decision, but one over which we have absolutely no control.

The chromosomes in those sperm cells and egg cells are not the exact same ones our mom and dad inherited from their parents. Bits and pieces of each one exchanged places while making their way into a sperm or egg cell, so that all of our chromosomes are mosaics of those present in previous generations, all the way back to our ancestors who lived in Africa fifty thousand years ago.

Variation in a single gene is sufficient to cause rare diseases such as phenylketonuria or Huntington's disease. Furthermore, variations in many genes contribute to our risk for common diseases such as cancer, heart disease, diabetes, Alzheimer's disease, Parkinson's disease, and many others. Tracking down the variants responsible for disease risk is now a major focus of medical research, and the pace of progress picks up every day. By the time you read this book, some people will know their own personal DNA code; we expect that you will know your own code sooner rather than later. When that happens you'll likely have a lot of questions.

We hope the basic concepts we've provided in this book will help you answer them.

We have tried to convey a few more ideas than those just summarized, notions about gene therapy, stem cells, pharmacogenomics, evolution, and race. But those were the dessert, and you may have taken in more calories than you need. If you're too full, remember that the concepts of genetics are simple, and they explain how small changes in your personal DNA code affect your health and happiness.

Notes and Further Reading

Chapter 1

The history of Google is told by David Vise and Mark Malseed in *The Google Story* (New York: Delacorte Press, 2005). Jim Watson has written several memoirs of his life in science, including *The Double Helix: A Personal Account of the Discovery of the Structure of DNA* (New York: Atheneum, 1968).

Chapter 2

Isaac Asimov told his story in *It's Been a Good Life* (edited by J. J. Asimov); Amherst, NY: Prometheus Books, 2002; we also used details from *Asimov: The Unauthorised Life* by Michael White (London: Millenium, an imprint of Orion Books, 1994). The structure of DNA and how it codes information, as well as the principles of basic genetic inheritance described in other chapters, is available in books online at the PubMed Web site <http://www.ncbi.nlm.nih.gov/sites/entrez>, especially *Molecular Biology of the Cell* by B. Alberts, A. Johnson, J. Lewis, M. Raff, K. Roberts, and P. Walter (New York: Garland Science, 2002); *Genomes* by T.A. Brown (New York: Garland Science, 2002); and *Molecular Cell Biology* by H. Lodish, A. Berk, S. L. Zipursky, P. Matsudaira, D. Baltimore, and J. E. Darnell, (New York: W. H. Freeman & Co., 1999).

Chapter 3

The story of how Patricia Stallings was falsely convicted of murdering her infant son has been told in several newspaper and magazine articles. We drew on articles by St. Louis Post-Dispatch reporters Bill McClellan ("Refusal To Accept Odd Coincidence Saved Stallings," St. Louis Post-Dispatch

September 25, 1991), Bill Smith ("Not Guilty! How the system failed Patricia Stallings," *St. Louis Post-Dispatch,* October 20, 1991), and Tom Uhlenbrock (St. Louis Post-Dispatch "Painfully true," March 18, 1993), by Rhonda Riglesberger ("How the Legal and Medical Systems failed Patricia and Ryan Stallings" in *Justice:Denied,* the magazine for the wrongly convicted, Volume 1, issue 8), and by Tim Graham ("When Good Science Goes Bad" in the October 2004 issue of *ChemMatters*). The case was described in the medical literature by James D. Shoemaker, Robert E. Lynch, Joseph W. Hoffman, and William S. Sly ("Misidentification of propionic acid as ethylene glycol in a patient with methylmalonic academia." *Journal of Pediatrics* 120:417–21). We were also informed by a report about the case that aired on December 9, 1998 on TLC Discovery Channel's Innocence Files show "Deadly Formula," and by conversations with Dr. William Sly of St. Louis University. Reliable information on genetic diseases can be found at many sites on the Internet, including the Mayo Clinic <http://www.mayoclinic.com>, the National Institutes of Health <http://health.nih.gov/category/GeneticsBirthDefects>, <http://www.genome.gov/27527652>, <http://www.genome.gov/11008303>, and their Genetics Home Reference Web site <http://ghr.nlm.nih.gov/>), and the National Organization for Rare Disorders <http://www.rarediseases.org/>.

Chapter 4

The history of the use of *Drosophila* in the study of developmental biology is told in *Fly: The Unsung Hero of 20th Century Science* by Brookes, M., New York: HarperCollins Publishers Inc., 2001; *Lords of the Fly* by Kohler, R.E., Chicago: The University of Chicago Press, 1994; and *Coming to Life: How Genes Drive Development* by Nüsslein-Volhard, C., Carlsbad, CA: Kales Press, Inc., 2006. E.F. Keller discussed the career of Christiane Nüsslein-Volhard in "*Drosophila* embryos as transitional objects: The work of Donald Poulson and Christiane Nüsslein-Volhard," *Historical Studies in the Physical and Biological Sciences.* Volume 26, Part 2, pp. 313–346, 1996.

Chapter 5

The story of the "Bubble Boy" David Phillip Vetter has been told many times; we relied mostly on an excellent article by Steve McVicker published in the

April 10, 1997 issue of the *Houston Press News* and *The Boy in the Bubble*, a documentary film produced and directed by Barak Goodman and John Maggio that aired on Public Television April 10, 2006. The use of gene therapy to correct SCID is described by Andrea Kon in the January 27, 2007, issue of *The Daily Telegraph*'s *Telegraph magazine* ("A chance for life"), by Gaspar et al. in *The Lancet* 364:2181–2187 ("Gene therapy of X-linked severe combined immunodeficiency by use of a pseudotyped gammaretroviral vector"), and by Hacein-Bey-Abina et al. in the *New England Journal of Medicine* 346:1185–1193 ("Sustained correction of x-linked Severe Combined Immunodeficiency by *ex vivo* gene therapy"). How gene therapy caused leukemia in some patients is explained by Hacein-Bey-Abina et al. in the *Journal of Clinical Investigation* 118:3132–3142 ("Insertional oncogenesis in 4 patients after retrovirus-mediated gene therapy of SCID-X1"). The story of Jesse Gelsinger's untimely death was told by Sheryl Gay Stolberg in the November 29, 1999, issue of the *New York Times* ("The Biotech Death of Jesse Gelsinger"). More information on gene therapy can be obtained from the American Society of Gene & Cell Therapy <http://www.asgt.org/>.

Chapter 6

The history of Elizabeth Hughes and the early days of insulin research is recounted in *The Discovery of Insulin* by M. Bliss (Chicago: University of Chicago Press, 1984). Information on the genetics of diabetes is available from the American Diabetes Association <http://www.diabetes.org/>. Information about stem cells and links to related resources can be found at the National Institutes of Health Stem Cell Information Web site (http://stemcells.nih.gov/).

Chapter 7

Pearl S. Buck provides an account of her daughter's life in *The Child Who Never Grew* (New York: The John Day Company, 1950). The medical history of phenylketonuria is told by S.E. Christ, ("Asbjørn Følling and the discovery of phenylketonuria," *Journal of the History of the Neurosciences* 12 [2003]: 44–54), and by S.A. Centerwall and W.E. Centerwall ("The discovery of phenylketonuria: the story of a young couple, two retarded children, and a scientist," *Pediatrics* 105 [2000]: 89–103).

Chapter 8

The story of the O'Brien's ill-fated Everest expedition is available at http://www.everestnews.com/ and news articles in the *Chicago Sun-Times*, May 4, 2005, by Lori Rackl; the *Oswego Post Standard*, May 2, 2005, by Mike McAndrew; and the *Seattle Times*, May 3, 2005, by Sara Jean Green. Alice Wexler describes her family's history with Huntington's disease in *Mapping Fate: A Memoir of Family, Risk, and Genetic Research* (Berkeley: University of California Press, 1995). The Hereditary Disease Foundation <http://www.hdfoundation.org/home.php> has extensive information about the current state of Huntington's disease research.

Chapter 9

Rita Hayworth's life is described in *If This Was Happiness: A Biography of Rita Hayworth*, by Barbara Leaming (New York: Viking, 1989) and *Rita Hayworth: The Time, the Place, the Woman* by John Kobal (New York: W.W. Norton and Co., 1977). Current research on Alzheimer's disease is available from the Alzheimer's Association <http://www.alz.org/index.asp>.

Chapter 10

The three examples of behavioral traits we used to illustrate the concept of heritability were drawn from the medical and scientific literature: K.L. Toh et al., "An hPer2 phosphorylation site mutation in familial advanced sleep phase syndrome," *Science*, volume 291, pp 1040–3; C.R. Jones et al., "Familial advanced sleep-phase syndrome: A short-period circadian rhythm variant in humans," *Nature Medicine* 5:1062–1065; Y. Xu et al., "Functional consequences of a CKIdelta mutation causing familial advanced sleep phase syndrome," *Nature* 434:640–644; P. McGuffin and D. Mawson, "Obsessive-compulsive neurosis: two identical twin pairs," *British Journal of Psychiatry* 137:285–287; R. Muhle, S.V. Trentacoste, and I. Rapin, "The genetics of autism," *Pediatrics* 113: 472–486.

Chapter 11

We learned about Aldred Scott Warthin and Pauline Gross from articles by Henry T. Lynch ("Aldred Scott Warthin, M.D., Ph.D. (1866–1931)" in *CA:*

A Cancer Journal for Clinicians 35:345–347), by Claudia Kalb ("Peering into the Future: Genetic Testing" in *Newsweek,* December 11, 2006), an unattributed article in the August 10, 1936 issue of *Time* ("G's Family"), and a radio show, "Daughter of Family G," written and produced by Ami McKay for the Canadian Broadcasting Corporation (available at www.soundprint.org). Much has been written about Katie Couric and Jay Monahan; two articles that we read are by Jacquelyn Mitchard (August 13, 2006, issue of *Parade* magazine), and Joanna Powell's interview of Katie Couric published in the October 1998 issue of *Good Housekeeping.* More information on colon cancer can be obtained from the Jay Monahan Center for Gastrointestinal Health <http://monahancenter.org/>. Information on all kinds of cancers can be obtained from the American Cancer Society <http://www.cancer.org/docroot/home/index.aspc> and from the National Cancer Institute <http://www.cancer.gov/>.

Chapter 12

Seymour Benzer reviews his life in interviews by Heidi Aspaturian, September 11, 1990-February 1991 as part of the Oral History Project, California Institute of Technology Archives. Pasadena, CA, <http://resolver.caltech.edu/CaltechOH_Benzer_S>. He is also the subject of a biography, *Time, Love, Memory* by J. Weiner (New York: Alfred A. Knopf, 1999). Benzer describes his early phage work in "The fine structure of the gene," *Scientific American,* January 1962.

Chapter 13

We learned much about all forms of Amyotrophic Lateral Sclerosis from the ALS Division of the Muscular Dystrophy Association <http://www.als-mda.org>. Especially helpful for our story were articles in their newsletter ("Mattingly Family Featured in Life Magazine," *ALS Newsletter* 3, no. 4 (October 1998); "Gene Mapped for Early-Onset, Slowly Progressive Form of ALS," *ALS Newsletter* 3, no. 2 (April 1998); "Gene Found for Early-Onset ALS" by Margaret Wahl, MDA/ALS Newsmagazine, 9, no. 5 (May 2004). The mapping and identification of the gene responsible for the Mattingly clan's ALS was described in articles by Phillip Chance et al. published in the *American Journal of Human Genetics* 62:633–640, "Linkage of the Gene for an Autosomal Dominant Form of Juvenile Amyotrophic Lateral Sclerosis to Chromosome 9q34"; "DNA/RNA Helicase Gene Mutations in a Form of Juvenile Amyotrophic Lateral Sclerosis (ALS4)" volume 74, pp 1128–1135).

Chapter 14

Henry Grunwald recounted his life in *One Man's America: A Journalist's Search for the Heart of His Country* (New York: Doubleday, 1997), and his battle with macular degeneration in *Twilight: Losing Sight, Gaining Insight*; New York: Alfred A. Knopf, 1999. The NIH <http://www.nei.nih.gov/health/maculardegen/armd_facts.asp> and the American Macular Degeneration Foundation <http://www.macular.org/> provide information about macular degeneration.

Chapter 15

The story of warfarin's development is told by K.P. Link ("The Discovery of Dicumarol and Its Sequels," *Circulation* 19 [1959]:97–107), and by R.L. Mueller and S. Scheidt ("History of drugs for thrombotic disease. Discovery, development, and directions for the future," *Circulation* 89 [1994]:432–449).

Chapter 16

The story of Alfred Russel Wallace and his contributions to evolutionary theory is well-told by Arnold C. Brackman in *A delicate arrangement: The strange case of Charles Darwin and Alfred Russel Wallace* (New York: Times Books, 1980), and by Wallace himself in *My Life*, volumes 1 and 2 (Chapman & Hall, LD, London, available at Google books). The full text of Darwin's and Wallace's first presentation of the theory of evolution by natural selection can be found at <http://linnean.org/index.php?id=380>. Much useful information about evolution can be found at The National Center for Science Education <http://ncse.com/>.

Chapter 17

Arthur Ashe's life is told in *Charging the Net: A History of Blacks in Tennis from Althea Gibson and Arthur Ashe to the Williams Sisters, by* C. Harris and

L. Kyle-DeBose (Chicago: Ivan R. Dee, 2007). A discussion of race and genetics is provided by R.A. Kittles and K.M. Weiss in "Race, ancestry, and genes: implications for defining disease risk," *Annual Review of Genomics and Human Genetics* volume 4, pp 33–67, 2003. The BiDil story is recounted in "The use of race and ethnicity in medicine: lessons from the African-American Heart Failure Trial," by J. N. Cohn, *Journal of Law, Medicine and Ethics* 34 (2006): 552–554.

Index